SAFETY CULTURE: THEORY, METHOD AND IMPROVEMENT

T0264884

Safety Culture: Theory, Method and Improvement

STIAN ANTONSEN
NTNU Social Research Ltd, Norway

CRC Press
Taylor & Francis Group
Boca Raton London New York

CRC Press is an imprint of the
Taylor & Francis Group, an **informa** business

CRC Press
Taylor & Francis Group
6000 Broken Sound Parkway NW, Suite 300
Boca Raton, FL 33487-2742

First issued in paperback 2017

© 2009 by Stian Antonsen
CRC Press is an imprint of Taylor & Francis Group, an Informa business

No claim to original U.S. Government works

Version Date: 20160226

ISBN 13: 978-0-7546-7695-9 (hbk)
ISBN 13: 978-1-138-07533-7 (pbk)

Visit the Taylor & Francis Web site at
http://www.taylorandfrancis.com

and the CRC Press Web site at
http://www.crcpress.com

Contents

List of Figures

List of Tables

Foreword

This book is devoted to studying the relationship between culture and safety.

I am indebted to a number of people who have contributed to my work in a variety of ways. First of all I wish to thank my boss, Professor Per Morten Schiefloe, for allowing me the time and resources to complete this book. I am also most grateful to Professor Jan Hovden. In addition to having read and commented on the manuscript, discussions with Jan have been an important source of inspiration for the ideas presented in this book. I would also like to thank Guy Loft at Ashgate for guiding me through the publishing process.

All my colleagues at Studio Apertura also deserve praise, in particular Petter Almklov, Rolf Bye, Jørn Fenstad, Trond Kongsvik, Lone Sletbakk Ramstad and Kristin Mauseth Vikland who have contributed to this book by engaging in numerous discussions about culture and safety.

Last, but not least, I wish to express my love and gratitude to my family for being patient and supportive throughout the process of completing this book.

Chapter 1

Introduction

On the 26 April 1986 Reactor Four at the Chernobyl nuclear power plant exploded, causing radioactive fallout over most of North-Western Europe. In many ways the Chernobyl accident changed the world by demonstrating the catastrophic risks involved in the most advanced technology humans have ever created. Although the disaster occurred in a high-tech system, its root causes were not all that high-tech. According to Reason (1987), the Chernobyl disaster appears in fact to have been mainly due to human action. In investigating the disaster, the International Atomic Energy Agency (IAEA) identified a 'poor safety culture' at the plant and in Soviet society at large as the root cause of the accident (IAEA 1986). According to the investigation, both the Chernobyl plant and its institutional context were characterized by a culture that had become blind towards the hazards inherent in nuclear technology. Several accident investigations post-Chernobyl have pointed to culture as a key causal factor in the creation of accidents,e.g. Piper Alpha, Ladbroke Grove, Kings Cross, *Challenger* and *Columbia*.

The proposed relationship between organizational culture and safety is the topic of this book. This relationship, epitomized by the concept of safety culture, has undoubtedly become one of the hottest topics of both safety research and practical efforts to improve safety. For instance, most oil companies today have programmes devoted to improving the company's safety culture. Lessons learned from these accidents have also started to influence the statutory frameworks that dictate the safety management systems of organizations.[1]

Alongside a growing interest in the concept of safety culture, there is still considerable confusion as to what the concept actually means. Within the academic discourse on safety culture several attempts have been made to define and specify its nature and content. For instance, journals like *Work & Stress* and *Safety Science* have devoted special issues to the subject.[2] But between the various approaches to the study of safety culture there are still few common elements. Put bluntly, the only thing about which there seems to be some agreement is the need for further research (Hale 2000). James Reason underlines the same point when he states 'Few phrases occur more frequently in discussions about hazardous technologies than safety culture. Few things are so sought after and yet so little understood' (Reason 1997: 191). Rosness (2001) voices a similar criticism in a paper titled 'Safety culture: Just another buzzword to hide our confusion?'

---

1   For example, the Norwegian petroleum legislation.
2   *Work & Stress,* 12, 1998, and *Safety Science*, 34, 2000.

What these critics are pointing to is a lack of theoretical models and frameworks to explain the relationships between organizational culture and safety. The existing research on safety culture thus rests on an assumption that there is in fact a relationship between culture and safety but without this relationship having been subject to discussion and empirical investigation.

This forms the point of departure for this book. The aim of the book is to take a step back and shed light on some fundamental theoretical, methodological and empirical questions regarding the relationship between organizational culture and safety. It addresses the following general problem: *How can a cultural approach contribute to the assessment, description and improvement of safety conditions in organizations?* The term 'cultural approach' means looking at the way cultural processes and traits within organizations influence safety.

This general question encapsulates four subordinate questions:

1. What are the theoretical foundations of a cultural approach to safety? This relates to the specification of the concepts and analytical models involved in studying the relationship between organizational culture and safety and I shall discuss it in Chapters 2–4.
2. How can the relationship between organizational culture and safety be investigated empirically? This is a methodical question regarding the assessment of safety culture and will be treated in Chapter 5.
3. In actual organizations what links exist between organizational culture and safety? This is an empirical question regarding the way the culture of an organization influences safety within it and will be answered through the empirical analysis presented in Chapter 6.
4. How does a cultural approach contribute to improved safety? How can research on safety culture be translated into techniques and principles for improving safety? This issue will be discussed in Chapter 7.

Posing and answering these questions has three main objectives. First, the aim is to contribute to *theoretical development* regarding the relationship between organizational culture and safety. Each of the four questions targets what I perceive to be gaps in the existing research on safety culture. The analyses presented aim to increase our knowledge of how cultural traits and processes may influence safety in organizations. Second, the analyses have a *sociological ambition* in shedding further light on how informal aspects of work and organizing relate to the formal structures of organizations and how the match between formal and informal organization may influence safety. In this respect the book continues the tradition from industrial sociology exemplified by Crozier's *The Bureaucratic Phenomenon* (1964) and Lysgaard's *The Workers' Collective* (2001). Third, the analyses all aim to be useful in the development of *practical measures* for the improvement of safety.

In the remainder of this chapter I will briefly introduce the key concepts of culture and safety. In Chapters 2 and 3 I turn to the theoretical framework of safety culture. The presentation of theory and previous research is divided into three constituent parts. In Chapter 2 I will give an overview of how issues of organizational safety have been dealt with historically and a review of some existing research on safety culture. This includes a discussion of some of the weaknesses in the existing research on safety culture. In Chapter 3 I will review some key theoretical perspectives on organizational culture. This is important in order to clarify some theoretical foundations related to the concept of safety culture. Chapter 4 gives an account of the relationship between safety culture and the issue of power. Chapter 5 discusses the challenges inherent in the assessment of safety culture, in particular the limits to survey methods in safety culture assessment. Chapter 6 presents a case study of the relationship between culture and safety in a high risk environment. In Chapter 7 I turn to the way a cultural approach can be useful in improving safety, while Chapter 8 extracts some of the conclusions of the book.

What is (Organizational) Culture?

What is culture? This is probably among the most complex and most debated questions of the social sciences. As Kringen (forthcoming) has noted, addressing the concept of culture is like opening a Pandora's box unleashing most social science concepts and, as a consequence, a host of analytical and definitional issues. A comprehensive discussion of the conceptual labyrinths of culture lies outside the scope of this book as it would undoubtedly require at least a book of its own. In this section I will only give a brief and admittedly rather superficial account of the concept of culture. A more detailed account of the theoretical considerations regarding the concepts of culture and organizational culture is presented in Chapters 2 and 3.

The word 'culture' stems from the Latin *colere*, which means to grow or to process (Eriksen 1998). The study of culture has to do with those aspects of human life that are not aspects of biology or unprocessed physical environment. This is culture in its broadest definition; everything that is not nature is to be seen as culture. Within this broad concept anthropologists and sociologists have tried to specify more analytical definitions of culture. This has resulted in a plethora of different definitions; a study by the American anthropologists Kroeber and Kluckhohn (1963) found more than 160 different definitions of the word. Although this figure refers to the number of formal definitions and not necessarily differences in the understanding of what constitutes cultural phenomena,[3] their study illustrates that culture is a term used in a wide variety of meanings and contexts.

---

3   Formal definitions of culture usually do little to clarify how researchers conceive of culture (LeVine 1984). The way one conceives of culture is more visible through the phenomena that are studied, the methods used and the inferences made, rather than the words used to define the subject matter.

Within sociology, the concept usually refers to the values that the members of a group share, the norms they follow and the material objects they create (Giddens 1994). Social anthropologists tend to prefer somewhat wider definitions, ranging from culture as 'the complex whole' of knowledge, beliefs, ethics and customs (Tylor 1968, in Eriksen 1998) to the 'webs of significance' that humans both create and find themselves suspended in (Geertz 1973). The notion of culture deployed here lies somewhere between the sociological and social anthropological conceptions. I see culture as the frames of reference through which information, symbols and behaviour are interpreted and the conventions for behaviour, interaction and communication are generated. On the one hand culture is individual, as it provides the necessary frames through which action becomes meaningful. On the other, culture is public, as it consists of patterns of meaning that are shared by the members of a cultural unit but which nevertheless exist independently of any individual actor. This means that I view culture as both a cognitive and a relational symbolic phenomenon.[4] Aspects like language, symbols, explicit and tacit knowledge, skills, identity, values, assumptions, and the 'dos' and 'don'ts' of social life are all regarded as central to the study of culture. The aim of a cultural study is, as I see it, to seek an understanding of why different information, symbols, actions and forms of behaviour and interaction stand out as meaningful to the members of a community.

Organizations constitute the cultural contexts studied in this book. The relationship between the concept of organizational culture and the more general concept of culture needs some clarification. The anthropological definitions of culture are developed to cover cultural phenomena like nations or tribes. Culture in this sense is associated with communities where the primary socialization of its members takes place within the borders of the cultural unit. This is not the case with organizations. This means that organizational culture is not as deeply rooted in the members of the cultural unit as are the frames of reference and behaviour conventions of a nation or a tribe. In other words, the concept of organizational culture refers to cultural phenomena of a different order (Schiefloe 2003). Although aspects of organizational culture may be taken for granted by the members of an organization, the members will be likely to have a greater degree of reflexivity towards the organization's cultural frames than is the case with tribal or national cultures.

I reserve the term organizational culture to apply to the informal aspects of organizations. This means that I define the organization's structural arrangements as being outside the concept of organizational culture, although strictly speaking, they too are products of social construction. The reason for this is that the analytical separation between culture and organizational structures allows for analyses of the relationship between the informal and formal aspects of work and organizing. As I will discuss further in Chapter 6, this relationship is sometimes problematic. Moreover, the match between informal and formal aspects of work and organizing

---

4   See Shore (1996) for a comprehensive discussion of this dualism.

is one of the instances where culture may have an influence on safety, hence the need for an analytical distinction between the two.

One important clarification needs to be made about the way the term 'organizational culture' is used in this book. Labelling a culture 'organizational' does *not* imply that it is necessarily attributable to an organization as a whole. On the contrary, organizations, depending on size and complexity, usually consist of multiple cultures associated with different departments, hierarchical layers, occupations and so on. For instance, the community of seamen studied in Chapter 6 does not correspond to the boundaries of a formal organization. Nevertheless, the seamen share some models of how a seaman should relate to his work, his colleagues and superiors and to those outside the community, which may be labelled as an occupational culture (Van Maanen and Barley 1984). Strictly speaking, then, one should perhaps speak of 'cultures in organizations' rather than 'organizational cultures'. Mainly for reasons of simplicity, however, I choose to stick with the established concept of organizational culture. As I will discuss further in Chapter 3, the theoretical explications involved in the use (or misuse) of the concept of organizational culture are more important than the concept itself. In any case, the various labels one could put on cultural phenomena in organizations are all related to *work*, which usually takes place in organizations. This is something that justifies the use of the term 'organizational culture' as an umbrella concept for the various guises of culture in organizations, including occupational or professional culture.

This sketch of the concept of culture, while hardly doing justice to the width of theory and research on the topic, gives a provisional account of the way culture is conceived in the analyses to be presented. The second key concept that needs some clarification is the concept of safety. The next sections present an outline of the concept of safety and the way safety relates to the concept of risk.

What is Safety?

From Risk to Safety

Safety must be understood in relation to the presence of some hazard or risk. As a simplification, it is common to define risk as a function of the likelihood of an event occurring and the degree of seriousness of the consequences of that event. When the level of risk is low the level of safety is considered to be high and vice versa. However, the concepts of risk and safety are somewhat more complicated than this and a more thorough examination is needed in order to clarify the contents of the concepts and the boundaries between them.

Traditionally, the idea of risk was closely related to the forces of nature, like earthquakes, storms, volcanic eruptions and the like (Lupton 1999). In this sense, the concept of risk refers to events that are 'acts of God' in the sense that they are beyond human control. Over the last few decades, however, the concept of risk seems to have changed and its contents widened (ibid.). Today, the concept refers both to hazards created by humans and those created by nature.

Within sociology, Ulrich Beck's *Risk Society* (1992) has been influential in putting the concept of risk on the agenda. In it, Beck shows how the modern industrial production of welfare has had unintended consequences. In addition to creating welfare in Western societies we have also produced new sources of risk. Beck concludes that the concept of risk has become a root metaphor for late-modern societies. While risk was previously related to a *lack* of knowledge, the creation of new risks in late-modern societies is seen as a result of the *high* degree of knowledge that humans possess and utilize to change their environmental surroundings. The consequences of climate change, which have started to make their presence felt in the last few decades, are a typical example of such man-made risks.

In parallel with the increased public awareness of risk in Western societies, quantitative risk analysis (QRA) has emerged as a specialized field of research. The main focus of risk analysis is to support decision making by assessing and quantifying the risks associated with the operation and design of technical systems (Aven and Kristensen 2005). A risk analysis basically consists of a description of what may go wrong, how likely it is that something will in fact go wrong, and the consequences involved if these things go wrong. In the traditional approach to risk analysis risk is seen as something that exists objectively and the risk analysis as reflecting a true state of affairs with regard to risk. This objectivist stance has been criticized by proponents of a more Bayesian thinking. In a Bayesian perspective, risk is seen as a matter of subjective judgment on behalf of the risk analyst, not as a measurement of objective facts (Aven and Kristensen 2005). Although the resulting risk matrix is vital input to decisions on how to deal with the risks identified, the main emphasis of quantitative risk analysis is to identify and describe risks. This concept of risk views accidents as statistical events in the sense that their occurrence is assumed to be stochastic and random (Hopkins 2005). There is less focus on finding the root causes of accidents.

How one can and should determine risks has always been a controversial issue. The traditional quantitative risk analysis is based on the assumption that there is some objective and true level of risk 'out there' and that one can come close to estimating this through the use of standardized techniques. Cultural theorists like Mary Douglas and Douglas Wildavsky (1982) have voiced strong objections to this concept of risk. Their argument is that risk will always be, at least to some extent, socially constructed. Risk will always be rooted in a social context and will be influenced by social processes and cultural patterns (Hovden and Larsson 1987). What is regarded as dangerous will therefore vary across cultural contexts. Consequently, there can be no objective, universally true, measure of risk. For instance, research has shown that people are usually more afraid of events that in all likelihood they will never experience, such as nuclear radiation and plane crashes, than the events that are quite likely to cause them serious harm such as driving a car or painting their house (cf. Tversky and Kahneman 1974; Slovic et al. 1980; Hviid Nielsen 1994). Douglas and Wildavsky's point is, of course, not to deny the existence of dangers. Douglas herself stresses that the 'dangers are only too real' (Douglas 1992: 29) but that our *evaluation* of them, that is the risk that we attribute to dangers, is contingent upon social and cultural context.[5] Decisions regarding what constitute acceptable risks and the definition of risk acceptance criteria thus have a heavy cultural component (cf. Fischhoff et al. 1981).

If the concept of risk is related to identifying dangers and estimating the likelihood of their occurrence, although this will never be an objective measure, the concept of safety refers to our ability to handle or control these dangers. Consequently, safety has to do with minimizing risks. This underlines the complementarity of the concepts. It also points to the fact that the concept of safety refers to the measures taken to minimize risks, either by reducing the probability of a hazardous event occurring or by reducing the consequences of the event if it does.

Based on the above delineation of the concepts of risk and safety, a definition of safety will consist of three elements:

1. The concept of safety refers to a *state* or *situation* where the statistical risk is deemed to be acceptable or as low as reasonably practicable, the so-called ALARP principle, (cf. Reason 1997).
2. Safety refers to a *feeling* of security and control. This feeling may or may not resonate with statistical descriptions of risk. Someone's feeling of being safe and secure is much related to the degree of trust in safety systems and public institutions (Drottz-Sjøberg 2003).
3. Safety constitutes a form of *practice* in the sense that it refers to our ability to reduce or eliminate the likelihood of hazardous events occurring. Practice here refers both to aspects of work performance and to the barriers (physical, organizational or technological) which serve to reduce the likelihood of accidents occurring and/or to limit the consequences of accidents that do occur.

This latter part of the definition underlines that risk and safety are not synonymous concepts as safety is related to improvement rather than description. Within the safety discourse accidents do not simply 'occur', they are always seen as *caused* by something (Hopkins 2005). This means that safety research aims at analyzing *why* accidents happen not only *how often* they may occur. Thus, while they are certainly inextricably connected, risk research and safety research have a somewhat different starting point.

There seems to have been a shift in research focus regarding risk and safety. The interest in technological, human and organizational factors that might increase

---

5   In an attempt to bridge the gap between 'positivist' and 'subjectivist' views on risk assessment, Shrader-Frechette (1991) has introduced the concept of scientific proceduralism to denote a middle position.

our ability to deal with risk is increasing (Gherardi and Nicolini 1998; Pidgeon and O'Leary 2000). This does not mean that the research on safety is an entirely new field of research or that risk researchers are a dying breed. Rather, I see it as an indication of a change in the balance between the descriptive risk research and the more applied safety research, with the latter becoming increasingly important.

Safety Against What and for Whom?

The concept of safety always includes a notion of hazard. Whether one is talking about safety as a state/situation, a feeling or a practice, we are always talking about being safe against or from something. Thus, when talking and writing about safety there will always be a need to specify the types of incidents involved.

This book focuses on safety related to work organizations. This means that risks related to natural disasters are not included in the analysis. The same goes for risks related to malevolent acts such as sabotage or terrorism. When the term 'safety' is used it pertains to safety against unintended accidental events that occur in work organizations. This type of event can be further specified into major accidents, involving a catastrophic potential with regard to human, environmental and/or financial losses, occupational accidents, involving few persons and less severe financial losses and third-person injuries, to passengers, road-users etc. In the cases studied, the risks involved are related to major accidents or occupational accidents involving personnel working on offshore service vessels, supply bases and offshore installations. The activities of the oil and gas industry also involve the risks of severe environmental damage. Third-person injuries are not part of the risk picture of the organizations studied here and are consequently not part of the concept of safety employed in the analysis.

Chapter 2
Culture and Safety

Safety in Organizations

The methods employed to improve safety vary greatly. Hale and Hovden (1998) divide the history of safety research and safety improvement into three phases. In the first phase, safety was seen as a technological issue and improvement was sought by developing safer machines and equipment. This phase lasted from the 1800s to the mid 1900s (ibid.). The second phase was characterized by an increasing focus on improving safety through strategic recruitment, upgrading employees' skills and making efforts to increase employee motivation. These are all measures aiming at improving work at the individual level.

The third phase of safety research starts around 1980 and is characterized by an increased focus on the organizational conditions for safety, especially the role of management systems. The so-called 'safety management philosophy' is, to a large extent, based on American management theory and on the assumption that accidents are mainly caused by human error or failure (Haukelid 1999). According to this reasoning, a safe organization is designed by creating a management system that specifies objectives, distributes responsibility, plans, organizes and controls according to safety precautions. In this perspective there is no big difference between safety and other aspects of the organization and safety can consequently be managed through the same principles as any other organizational function (ibid.).

Since safety culture has to do with the organizational conditions for safety, the cultural approach is in many ways an extension of the third phase of safety research described by Hale and Hovden. There are, nevertheless, some important differences between the safety culture approach and the safety management philosophy. The safety management philosophy is predominantly oriented towards *formal* organization. The safety culture approach, on the other hand, is oriented towards the *informal* aspects of the organization.

The increased interest in the relationship between culture and safety is probably associated with a more general shift in the way we think about how organizations are created and how they function. The classical view of organizations rests on an assumption that organizations and organizational members act according to strictly rational principles, where all problems are analysed and solved by deliberate calculation of which action strategies are the most likely to give maximum benefit. This assumption has come under increasing criticism over the last thirty years. Several theorists have stressed that actions are just as much based in non-rational aspects as purposive calculation (e.g. Brunsson 2000).

Norms and beliefs, among other factors, can give quite detailed guidance for action and behaviour and thereby bring about established action patterns which are not necessarily in accordance with a rational model for action.

This reasoning is probably also involved in the criticism which some culturally oriented writers direct at the belief in the potential for controlling behaviour through rules and procedures inherent in the safety management philosophy (e.g. ACSNI 1993).

Having located the research on safety culture within the historical landscape of safety research, it is now time to move on to describing the concept of safety culture and the way it has been analysed in the existing research on the topic.

Safety Culture: Background, Theory and Research

Conceptual Background – The Chernobyl Accident

The investigation report following the Chernobyl accident in 1986 is usually cited as the first publication to use the term 'safety culture'. In another report on Chernobyl, the International Atomic Energy Agency even emphasizes culture as the *main* root cause of the explosion in reactor four:

> The accident can be said to have flowed from deficient safety culture, not only at
> the Chernobyl plant, but throughout the Soviet design, operating and regulatory
> organizations for nuclear power that existed at the time. (IAEA 1992: 23)

Among the traits that in sum constituted the 'deficient safety culture' identified by the IAEA, were a reluctance to question the decisions of one's superiors, a propensity for procedural violation – the alarm systems were shut down during the tests that sparked off the accident sequence – a complacent belief in the ability to control the technology and making the production of energy, not the upholding of safety, the key priority of the plant's managers and operators (Reason 1987; Medvedev 1991).

In many ways, the Chernobyl disaster represented a crossroads in safety research and the way organizations go about improving safety, in that it once and for all demonstrated the influence of the human element on risk and safety (Meshkati 1996). It left little doubt that 'soft' aspects of organization, like culture, can indeed have very 'hard' consequences. Today, more than 22 years after the accident, there is still a zone of 3 kilometres around the plant that is classified as uninhabitable due to the high levels of radiation. The neighbouring town of Pripyat, which once housed 50,000 people, is still completely abandoned.

Two years later, the oil platform Piper Alpha exploded in a fireball in the British part of the North Sea. Here, too, the investigation concluded that cultural issues, both on the platform and in the oil company in general, had contributed strongly to creating the accident. Among the concluding remarks of his report, Lord Cullen stresses that: 'It is essential to create a corporate atmosphere or culture in which safety is understood to be and is accepted as, the number one priority' (Cullen 1990: 300).

In the years following Chernobyl and Piper Alpha, interest in the potential safety consequences of culture made its way into the community of safety practitioners. Several business organizations and regulatory authorities began publishing guides and manuals aiming to help safety managers create 'good safety cultures' in their companies, a development pioneered, not surprisingly, by the nuclear industry. Within this literature, a host of organizational qualities were assumed to be the key components of a 'good safety culture', for instance good communication and a high level of employee competence (see, in particular, IAEA 1991 and ACSNI 1993). The empirical basis for the qualities selected is rarely discussed and there is a clear tendency towards focusing on the organizational dimensions that are the easiest to measure and monitor (Cox and Flin 1998). This type of publication and a great deal of the interest in safety culture in general, I suspect, stems from an assumption that culture can be modified or engineered in order to improve safety in high risk organizations. This assumption is far from unproblematic as I shall discuss further, below.

Safety Culture – Some Theoretical Perspectives

Prior to and parallel with the concept of safety culture being introduced in accident investigations, some safety researchers with a theoretical inclination had already started to grapple with the issue of culture in relation to risk and safety.

Within the safety literature there are primarily two major theoretical frameworks of interest to scholars interested in safety culture. Barry Turner's *Man-Made Disasters* (1978) offers one of these. Turner was among the first to undertake a comparative study of the accident sequences of major disasters. The other strand of theorizing originates from a group of researchers at the University of California, Berkeley, and the University of Michigan. This group, with Karlene Roberts, Todd R. La Porte and Karl Weick among the key participants, has studied the common features of organizations that display exceptionally high levels of safety, and which are consequently labelled High Reliability Organizations (HROs). In the next section I will review the major contributions of these two strands of theory in relation to the study of safety culture.

Turner's Man-Made Disasters   *Man-Made Disasters* (Turner 1978; Turner and Pidgeon 1997) is undoubtedly the seminal work on the relationship between cultural processes and organizational safety. With this work Turner was among the first to conceptualize and study accidents as processes, not merely as sudden, unforeseen events.

Turner's model is based on what has been labelled a 'grounded theory' approach by Glaser and Strauss (1967). This is a qualitative approach aimed at the construction of theory on a basis of empirical material. In other words, it is primarily an inductive approach, although the 'grounding' of theory in data also involves a deductive component. His analysis is based on a systematic study of the official investigations into 84 accidents in the UK between 1965 and 1975. Grounded in these data, he constructs a generic model of accidents, showing that they rarely come as bolts from the blue. Rather, accidents are created through complex chains of events that accumulate over time and so the accident sequence is usually initiated long before the occurrence of the triggering event. This led Turner to draw an analogy between the build-up of accidents and the outbreak of a disease, since they both seem to progress through a phase of *incubation*. Within the incubation phase it is common to find instances of so-called 'decoy-phenomena' (Turner and Pidgeon 1997: 48), which denotes the presence of smaller hazards and problems that lead the organization's attention away from the 'real' ones. Turner describes the incubation phase in general as characterized by 'a range of many types of *information* and *communication difficulties* associated with the ill-structured problem which eventually generates the accident' (Turner 1992: 194). The incubation phase ends when the different latent factors coincide to create the event that triggers the disaster. This way of viewing accidents as created by the concurrence of latent conditions and active failures has been highly influential for later accident models, most notably James Reason's 'Swiss cheese model'(Reason 1997). Turner's accident model is shown in Figure 2.1.

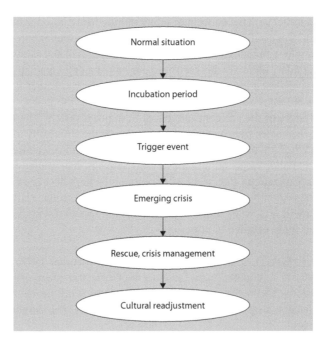

Figure 2.1   Turner's '*Man-Made Disaster*' model

Adapted from Turner (1992).

When accidents go through an incubation phase it implies that there will be signs or symptoms that things are about to go wrong *before* the onset of the crisis. For instance, Turner sees repeated violations of procedures as symptoms of an organization being in a state of incubating an accident. The ability of an organization to detect such danger signals is one of the aspects of Turner's framework that is of particular interest to those interested in safety culture. Turner specifies that the neglect or failure to detect such signals depends on what he labels 'rigidities of perception and beliefs' (Turner and Pidgeon 1997: 47). These perceptions and beliefs constitute the frames of reference through which organizational members relate to information, their activities and the organization's surroundings:

All organizations develop within them*selves*, as part of the equipment which they use in operating *in* the world, elements of continuous culture which relate to the tasks they face, to the environment they find themselves *in* and to the manner in which *the members of* the organization are to interact both with each other and with the equipment they may use. (Turner and Pidgeon 1997: 47)

Turner views this ability for cultural coordination as a key condition for the effectiveness of any organization. This very property, however, also involves the risk that something falls outside these frames of reference causing a 'collective blindness' towards specific hazards (Turner and Pidgeon 1997: 47).

The cultural frames of reference include a set of shared perceptions about what is to be considered safe and what dangerous. The more widely shared these cultural assumptions are, the less the organization's ability to detect signals of danger that fall outside these frames of reference. One of Turner's case studies illustrates this. In the village of Aberfan in South Wales, 144 people, 116 of whom were children, were killed when a colliery waste tip collapsed, slid down the mountainside and crushed twenty houses and two schools. Similar events had occurred several times at other mining locations in Britain and the local residents had protested against the possible danger from the tips at Aberfan. The mining industry, though, proved to be impervious to such warnings, in spite of the introduction of regulations for locations where colliery tips could be placed. Turner sees this as a consequence of the prevailing beliefs and assumptions of the industry being directed towards problems and difficulties related to the extraction of coal. The problem of disposing of the colliery waste was a mere by-product of the organization's core tasks and was thus pushed out of its 'field of vision' about risk. This illustrates an inherent paradox of culture, that it is at the same time a 'way of seeing' and a 'way of *not* seeing' (Turner and Pidgeon 1997: 49). This aspect of culture offers an explanation as to how it is possible for organizational members to ignore signals of danger that are obvious with hindsight and that perhaps were also visible at the time to people outside the organization.

Turner's insights are of pivotal importance to the study of safety culture and his later works deal with safety culture explicitly (Turner 1991, 1992; Pidgeon et al. 1991). Nonetheless, with the exception of a few studies (e.g. Vaughan 1996; Gherardi and Nicolini 2000), this work seems to have passed virtually unnoticed by much of the existing empirical research on the topic. I see this as a common shortcoming of a large part of existing empirical research on safety culture since Turner's model is a powerful way of conceptualizing the relationship between culture and safety.

Turner's analysis highlights some organizational conditions that may be involved in the *creation* of accidents. Partly inspired by Turner, some researchers on High Reliability Organizations have aimed to identify conditions that may remedy rigidities of beliefs and thus improve organizations' abilities to *prevent* accidents.

Some characteristics of High Reliability Organizations (HROs)   As indicated, studies of HROs seek to identify the properties of organizations characterized by excellent safety records. Although this research aims to identify success criteria rather than the conditions that foster failure, the methodological approach bears strong resemblances to the grounded theory approach employed by Turner.

The organizations classified and studied as HROs are usually complex systems such as aircraft carriers, nuclear submarines and nuclear power plants. These are systems that according to other safety theories, most notably Charles Perrow's (1984) 'Normal Accident Theory' (NAT) should have failed. The HRO researchers therefore voice some of their arguments directly in opposition to Perrow's more pessimistic view on the possibility of maintaining reliability in complex systems. It could be argued that some of the organizations studied within the HRO school, such as air traffic control and aircraft carriers (cf. Roberts 1993), are *not* among the type of organizations singled out by Perrow as prone to normal accidents (cf. Perrow 1994, 1999). Nevertheless, the dichotomy between HRO and NAT has become one of the major theoretical frontiers within safety research.

The literature on high reliability organizations emphasizes that the safety level of organizations is strongly influenced by the degree of overlap in operators' tasks and areas of expertise (organizational redundancy), the ability to reconfigure the organization from centralized to decentralized structures in times of crisis, and from formal to informal patterns of interaction (LaPorte and Consolini 1991). A high degree of redundancy allows for having more eyes and minds to assess the same processes and decisions and thus a higher capacity for error control. The ability for structural and interactional reconfiguration makes it possible to be both rule-based and reliable in times of normal operation and flexible and innovative in times of crisis.[1] These factors are, first and foremost, related to organizational design and organizational structures.

The HRO literature also emphasizes cultural conditions for safety. Karl Weick (Weick 1987; Weick et al. 1999) is the HRO theorist with the strongest emphasis on the role of cultural aspects in maintaining safety. He views culture as an important informal coordination mechanism which can constitute an alternative to

---

1   This ability for reconfiguration is the HRO theorists' solution to 'Perrow's dilemma' stating that organizations characterized by tight couplings and interactive complexity are bound to fail since they need to be both centralized and decentralized at the same time.

the structural centralization which Perrow (1984) sees as the only way to deal with the risks connected with extremely complex and tightly coupled organizations and technologies.

Weick's main argument is that organizations that value narratives and storytelling will be more reliable than other organizations because stories express important information about how the organization works, about events that may occur and how such events may be dealt with (Weick 1987). In this way, organizational culture may be a source for the creation of safety by being a medium for the communication of safety-critical knowledge. It may thus serve as an opportunity to learn *without* having to rely on trial and error.

In his early works, Weick seemed to overlook the ways emphasized by Turner (1978) in which culture may affect safety negatively. However, Weick develops his perspective in his later works (Weick et al. 1999; Weick and Sutcliffe 2007), where he fuses his earlier thinking with Turner's perspectives. In his recent works, Weick introduces the concept of 'organizational mindfulness', seen as a common feature of HROs. Organizations with a high degree of collective mindfulness are characterized by placing great value on learning and troubleshooting, by being sensitive towards the possible consequences of work operations and by valuing flexibility and expertise. The concept of mindfulness provides the opportunity, although implicitly, for conceptualizing the way culture may negatively influence safety.

Some of the research within the HRO school may be criticized for its somewhat biased selection of cases. It is especially hard to see how the organizational properties of naval aircraft carriers and other military units can be replicated in civil organizations. The research on HROs has nevertheless identified a number of 'ideal types' of safety-inducing organizational properties that have had significant influence on safety research.

The organizational characteristics emphasized by Weick and colleagues (1999) and those described by Turner (1978) seem to have a great deal in common. Both Turner and Weick stress that they are describing properties of social relations, not individual properties. Here there is a gap between the theoretical perspectives and the empirical research, differences that I will elaborate on further in the next section.

Existing Research on Safety Climate and Safety Culture

The research on safety culture is by no means characterized by conceptual consensus, as I have already indicated. On the contrary, the existing research appears fragmented and theoretically underspecified (Pidgeon 1998). Just as the concept of culture has become known for its plethora of definitions and conceptualizations, the concept of safety culture has been defined and employed in very different ways. The most frequently cited seems to be a definition from the

Advisory Committee on the Safety of Nuclear Installations (ACSNI). Here, safety culture is defined as:

> The product of individual and group values, attitudes, perceptions, competencies, and patterns of behaviour that determine the commitment to, and the style and proficiency of, an organisation's health and safety management. (ACSNI 1993: 23)

This definition forms the basis of much research (e.g. Lee 1996; Cox and Cheyne 2000; Lee and Harrison 2000). It is also the 'official' definition applied by the British Health and Safety Executive (HSE). However, this definition is almost identical to the earlier definition of safety *climate*, which is a conceptual predecessor of the concept of safety culture. The relationship between the two concepts has been subject to much discussion and represents a natural starting point for a description of safety culture research.

Safety Culture and Safety Climate

The relationship between safety climate and safety culture has been thoroughly discussed elsewhere (cf. Hovden 1991; Cox and Flin 1998; Guldenmund 2000). I will therefore confine myself to delineating the main features here.

Even though Guldenmund (2000) traces the research on safety climate back to the 1950s, it is Zohar's (1980) study of 20 manufacturing companies in Israel that is ubiquitously cited as the originator of the concept. Zohar's findings indicate that the safety levels of these organizations was influenced by managers' attitudes towards safety and the perceived priority given to safety training . Later research has confirmed much of Zohar's findings. A study by Flin et al. (2000) reports many of the same factors. They conducted a meta-study of previous empirical attempts to study safety climate and found five broad themes that pervade much of the research:

1. *Management*. Managers' commitment to safety in relation to other organizational goals, e.g. production, is the most recurrent theme in the study. This factor presumably pertains to the commitment of senior managers, middle managers and supervisors .
2. *Safety system*. Various aspects of the safety systems within organizations also constitute a recurrent theme in the studies reviewed by Flin and colleagues. Among the dimensions in this factor are respondents' opinions on safety policies, safety equipment and permit to work systems. Opinions related to accident and incident reporting could probably also be placed under this heading.
3. *Risk*. This category consists of perceptions and attitudes towards risk and safety, including risk-taking behaviour and perceptions of worksite hazards.

4. *Work pressure.* Issues related to work pace and workload must be seen in close relation to the management category and concern the balancing of safety and production. This is probably the most widely recognized component of safety culture, and one that may be increasingly important given the fierce competition and drive towards cost reduction that characterizes the global economy.
5. *Competence.* This factor encompasses aspects like selection and training of the work force, as well as the company's assessment of worker competence.

In addition to these five themes, Flin et al. also mention the role of procedures and rules. Although they do not find this a recurring theme in the studies reviewed, they emphasize workers' compliance with or violation of rules as a topic that may be worthy of more attention.

So, in terms of these themes, a 'good' safety climate is one where managers at all levels are highly committed to safety; where the workforce express satisfaction with and adherence to the organization's safety system; where everyone is risk averse; where there is no pressure towards maximizing profits at the expense of safety and where operators as well as managers are highly qualified and competent.

These dimensions of safety *climate* are commonly included in research on safety *culture*, and the two concepts are used interchangeably, although there is a clear trend that culture is gaining popularity at the expense of climate. The blurred borders between them are illustrated by the fact that the study of safety climate by Flin et al. is published in *Safety Science*'s special issue on safety culture.

Although the definitions and analytical dimensions of climate and culture are highly similar, there are important conceptual differences between the two. While culture is normally taken to refer to meanings and beliefs that are deeply rooted and often taken for granted, climate is seen as a more superficial manifestation of culture (Reichers and Schneider 1990; Schein 1992). In an attempt to clarify the relationship between the concepts, Cox and Flin (1998) have described culture as an organization's personality while climate is seen as the organization's mood. Culture is thus a concept on a higher level of abstraction signifying traits that are fairly stable over time. Climate, on the other hand, refers to more transient characteristics that are both more visible and easier to change than culture.

Within safety research there is, unfortunately, a tendency for this conceptual difference to be downplayed. The concept of culture seems to be in the process of replacing the concept of climate but this shift is often more a change in the words used to signify the same phenomena than a shift in the phenomena studied and the research approaches employed. The reason for this is probably that if climate is seen as a manifestation of culture, the thought appeals that it is sufficient to study climate to be able to draw inferences about culture. This is, however, a problematic strategy. As Schein (1992) has stressed, there is not necessarily a consistency between the espoused values (climate) and the basic assumptions that form the foundation of an organization's culture. With reference to Argyris and Schön's (1996) distinction between espoused theory and theory-in-use, Schein points out that there will often be a considerable discrepancy between what we *claim* to do and what we actually *do*. The basic assumptions that constitute the cultural core in Schein's model are expressed through the latter. According to Schein, the degree of consistency between climate and culture is an empirical question. Consequently, a study of climate cannot be regarded as sufficient empirical basis for making inferences about an organization's culture.

Despite this, as already indicated, there are no clear borders between the research on safety climate and safety culture. When I now move on to describe different approaches to safety culture, it is difficult to distinguish explicitly between studies of climate and studies of culture. I therefore choose to treat studies of safety climate and safety culture as one field of research.

Different Approaches to Safety Culture

Research on safety culture has traditionally been dominated by both psychological and engineering perspectives. Put somewhat simplistically, one might say that psychologists have primarily been preoccupied with studying attitudes and behaviour in relation to safety, while researchers from engineering disciplines have placed more emphasis on the safety systems that provide frames for behaviour in organizations (Pidgeon 1997, 1998).

Within the psychological research, so-called psychometric approaches form the majority of studies (Cox and Flin 1998). These studies aim to identify traits of safety cultures by conducting standardized questionnaires with large populations of employees. Sometimes these studies are repeated in the same populations as attempts to provide a regular monitoring of the safety conditions of an organization. Such studies are in practice to be classified as studies of safety climate. Continuing the tradition from Zohar (1980), they often rely on factor analysis or structural equation modelling (SEM) as techniques for uncovering the 'underlying' dimensions of a set of attitude variables.

The psychometric survey represents a fairly inexpensive research design and allows for the use of established techniques for analysis so it quickly became an area of growth within safety research (Cox and Flin 1998). Rundmo's (2000) study of attitudes and risk perception in Norsk Hydro Ltd is one example of a psychometric study. Huang and colleagues' (2006) study of safety climate among the employees in 18 American corporations is a more recent example. Other similar examples include studies by Lin et al. (2008), Håvold and Nesset (2009) and Flin (2007).

The research originating in engineering disciplines has a clearer emphasis on management and lines of responsibility than the psychological research. As indicated above, different properties of safety management systems were among the 'big five' of safety climate/culture dimensions identified by Flin et al. (2000). The investigation of such properties has previously been studied through safety audits and it is only in recent years that such audits also claim to be assessments of safety culture in addition to investigating the properties of safety systems (Cox and Flin 1998).[2] An example of this conceptualization of safety culture can be found in one of the International Atomic Energy Agency's leaflets on safety culture. Here, safety culture is defined as consisting of two components.

The first is the necessary framework within an organization and is the responsibility of the management hierarchy. The second is the attitude of staff at all levels in responding to and benefitting from the framework (IAEA 1991: 5).

The second definition, particularly, makes it clear that the engineering perspective on safety culture also includes a psychological aspect.

The main emphasis is nevertheless on formal organization, which constitutes the 'framework' referred to in the definition. This emphasis becomes particularly evident when considering the list of 'indicators' suggested for the assessment of safety culture in the same report. Of the 143 indicators proposed, around 100 relate to administrative procedures and safety policies. Examples of such indicators are questions as to whether documents describing formal lines of responsibility are updated and whether management has regular meetings with regulatory authorities (IAEA 1991). This illustrates that research within the engineering perspective is somewhat more oriented towards practical safety improvement than seems to be the case with psychological research.

Although research within this perspective was probably more common in the years after Chernobyl, there are still examples of similar conceptualizations of culture. Mitchison and Papadakis' (1999) and Duijms and Goossens' (2006) discussions on the 'measurement' of culture through general safety audits can be viewed as more recent representatives of this perspective. As I shall discuss in further detail, what is lacking in both the psychological and the engineering perspectives on safety culture is an understanding of the fact that culture is something that is socially *shared* among the members of a group or an organization.[3]

A third strand of research, that of human factors and ergonomics, may also be identified, although there are no clear boundaries between this research and the two previously mentioned 'schools' of research. Human factors research and engineering focus on how humans interact with their tasks, their physical workplace – especially the machines and other technology they use – and their social environment (cf. Wickens et al. 2004; Salvendy 2006). Although the labels human factors or ergonomics encompass a wide variety of research, there seems to be a preeminent focus on individual behaviour and human error (e.g. Reason 1990; Johnson 1996).

As well as the three perspectives already mentioned, there are now signs that some authors in this field are starting to link the research on safety culture more

---

2   The methodology involved in Shell's 'Hearts and Minds' programme is a good example of the fusion between safety audits and safety culture assessment (see Hudson et al. 2002).

3   The view that culture is shared by the members of a group is contested by some authors. This is discussed in detail in Chapter 3 of this book.

tightly to organizational theory and more general safety theory (e.g. Pidgeon 1997, 1998; Pidgeon and O'Leary 2000; Gherardi et al. 1998; Gherardi and Nicolini 2000; Guldenmund 2000, 2006; Rosness 2001; Langåker 2002; Richter and Koch 2004; Hopkins 2000, 2005, 2006). Many of these studies employ a concept of culture that is more inspired by anthropology than the three previously mentioned perspectives. Drawing on different theoretical traditions also involves differences in the purpose of the research. The early research on safety culture was normative, in the sense that it aimed directly at the improvement of safety. This normative aspect is somewhat less emphasized in the more anthropologically inspired studies cited above. Although the aim is still to contribute to the improvement of safety in the organizations studied, there seems to be a greater emphasis on making a thorough *description* of the organizations involved before turning to the question of improvement.

Despite this recent development, however, the mainstream of safety culture research still rests on a relatively unclear theoretical foundation. This is a serious shortcoming for at least two reasons. First, it of course increases the problems and confusion surrounding what safety culture is and how it relates to other aspects of work and organizing. The less the concept is related to a theoretical framework the more difficult it will prove to specify a precise analytical framework for the empirical analysis. Second, being only loosely coupled to theoretical models has made it difficult for the existing research to give satisfactory answers to questions of how culture may actually influence safety. Without making this coupling firmly it is hard to see why we should be interested in safety culture in the first place (Hale 2000).

Some Common Weaknesses in Many Studies of Safety Culture

As indicated in the previous sections, the loose couplings between the empirical studies of safety culture and the existing theoretical perspectives, most notably those of Turner (1978) and Weick et al. (1999), represent a common weakness in many studies. The lack of theoretical models explicating the relationship between organizational culture and safety has also been stressed by Frank Guldenmund (2000), in his meta-analysis of several studies of safety culture. Guldenmund does not mention the works of Turner and Weick as possible solutions to this problem. Rather, he sees the solution to the theoretical problems of safety culture research in a stronger coupling to the 'mother concept' of *organizational* culture. More specifically, he proposes an approach based on Schein's (1992) model of organizational culture, where culture is divided into different layers consisting of artifacts, espoused values and basic assumptions.

Guldenmund no doubt has a point when he draws attention to the weak links between safety culture research and organizational culture theory. For instance, it is quite striking that much of the discussion around the relationship between safety climate and safety culture has proceeded quite disconnected from the corresponding debate on organizational climate/culture from the mid 1980s onwards (cf. Schneider 1990).

Another indication that safety culture research has not utilized the existing knowledge on organizational culture is seen in the way publications on safety culture make reference to organizational theory. Schein's *Organizational Culture and Leadership* (1992) is something of a standard reference in journal articles on safety culture, but exactly *how* the studies relate to Schein's framework is rarely discussed. In fact, the link to Schein's framework is highly problematic in many studies of safety culture since Schein expresses a great deal of scepticism regarding psychometric studies of culture. For instance, he criticizes research on organizational climate for being too superficial to give valid accounts of culture:

> If one does not decipher the pattern of basic assumptions that may be operating, one will not know how to interpret the artefacts correctly or how much credence to give to the articulated values. (Schein 1992: 26)

This criticism could also be raised against studies of safety culture, many of which refer to Schein as the source of their conception of culture. This illustrates that the concept of culture has been incorporated into safety research in a rather uncritical and unprocessed way. That is, the idea that culture is important to understand how safety is created and maintained was brought into safety research but the theoretical perspectives that can tell us how, were left behind. This represents a problem since it leaves safety culture research to 'invent the wheel' all over again, instead of utilizing the insights already present within research and theory on organizational culture.

Another criticism of the early safety culture research is that it places too much emphasis on individual properties. As Pidgeon (1998) has stressed, culture, by definition, is something shared between the members of a group. Consequently, it is not sufficient to assess the attitudes and values of individuals and view the aggregate of these attitudes and values as the culture of the group studied. Alvesson has stressed:

> Culture is not primarily 'inside' people's heads, but somewhere 'between' the heads of a group of people where symbols and meanings are publicly expressed, e.g. in work group interactions, in board meetings but also in material objects. (Alvesson 2002: 4)

This view on culture is in line with Geertz's (1973) emphasis on the 'public' nature of culture. Since culture is inextricably linked to language and since language is by definition shared by actors interacting with one another, culture and meaning are invariably public. With Wittgenstein (1997), one could say that this theoretical position involves shifting the attention from the players of a language game to the rules of that game.

This position clearly has some methodological implications. As I will discuss further in Chapter 5, this means that the study of culture cannot rely on questionnaires alone. Since many aspects of culture belong to the sphere of social life, the study of culture will require more interactive probing.

In relation to this criticism, Silvia Gherardi and colleagues (Gherardi et al. 1998; Gherardi and Nicolini 2000) have argued that the study of safety culture should focus on the specific practices in which safety is created and learned. They argue that safety is not a separate object of knowledge. Rather, safety is closely related to practice, as safety is seen as having to do with the practical knowledge of how to work and how to behave in certain situations. This is in many ways similar to Rasmussen's (1997) arguments towards studying human behaviour in light of normal work processes, instead of comparing behaviour to predefined procedures in order to detect human errors. Gherardi's approach, however, introduces to the study of safety culture a more distinct focus on social processes and conventions for behaviour which has been largely absent in the psychological and engineering views on safety culture.

The most recent criticism of existing safety culture research comes from authors like Höpfl (1994) and Richter and Koch (2004, see also Richter 2001). They introduce a far more complex perspective on safety culture. One that is better able to account for differentiation between groups in an organization, the issue of ambiguity of cultural aspects, and the possible fragmentation of cultures. This represents a noteworthy attempt to bring in more advanced organizational theory to the study of safety culture.

The concepts of differentiation, ambiguity and fragmentation have been thoroughly debated within the literature on organizational culture, particularly through the writings of Joanne Martin and Debra Meyerson (e.g. Meyerson and Martin 1987; Martin 1992).

By introducing these issues into safety research, these authors make a very important point, which is the fact that the bulk of safety culture research rests on an implicit model of organizational and cultural *harmony*. Issues of conflict and cultural inconsistency have been largely neglected by both the psychological and engineering perspectives on safety culture. Therefore, Richter and Koch (2004: 704) describe the existing research on safety culture as representatives of a 'unitary, integrative and monolithic approach to culture'. Although they go too far in rejecting the integrative aspects of culture at the same time as they implicitly accept the integration perspective by defining culture as something that is shared, Richter and Koch (2004) nevertheless manage to elevate the discussion on safety culture to a theoretical level which has hitherto been in short supply. Their article also spawned an immediate discussion of the nature of safety culture through a series of letters to the editor in a subsequent issue of *Safety Science* that also included a contribution by Schein (2004). In Chapter 4 I delve further into some of the questions raised by Richter and Koch. In particular, I discuss the way existing research on safety culture has dealt with, or rather, *not* dealt with, issues of power.

Let me attempt to summarize this review of the perceived state-of-the art of safety culture research.

While the psychological and engineering perspectives on safety culture have made important contributions to the understanding of safety in organizations, one might question the degree to which some of this research fits into the term 'culture' as it has been conceptualized within organizational theory, sociology and anthropology. Nevertheless, there are signs of change in this respect. There seem to be a growing number of studies relying on the insights of research and theory on organizational culture. I see this as important progress and it is the aim of the present book to contribute to this progress by further elaborating the way safety culture research may utilize the insights of organizational culture. However, the relationship between the concepts of safety culture and organizational culture has yet to be clarified, and thus it requires some discussion.

Safety Culture and Organizational Culture

Most studies, explicitly or implicitly, view safety culture as a subset of organizational culture; as those parts of organizational culture that influence safety (e.g. Clarke, S. 1999; Parker et al. 2006). This relationship is usually elaborated by various properties which are assumed to constitute a 'good' safety culture (e.g. Olive et al. 2006). In my view, this way of dealing with the two concepts skips several important questions.

First of all, how is one to know which 'parts' of organizational cultures influence safety? I see this as an empirical question, not one that can be answered by referring to a priori dimensions of 'good' safety culture.

Second, the definition of safety culture as a subset of organizational culture consisting of a specified set of dimensions involves a reification of culture, i.e. it reduces culture to a fixed entity with stable characteristics (Guldenmund 2000; Hale 2000). For instance, the definition cited above, defining safety culture as the 'product of individual and group values' (ACSNI 1993: 23), clearly treats culture as an entity. Many anthropologists would take issue with such conceptualizations of culture. Geertz (1973: 11), for instance, has voiced strong arguments against viewing culture as a 'self-contained "super-organic" reality with forces and purposes of its own', and proposes a more semiotic approach directed at uncovering the *meaning* conveyed through social action. Within anthropology, some have also argued that one should speak of the *cultural* rather than culture, i.e. using the word culture as an adjective rather than a noun (Keesing 1994; Haukelid 2008). While this may avoid some of the reification associated with culture as a noun, I suspect that many of the problems regarding reification would still be present. For one, language is in itself a reifying mechanism; attaching a linguistic sign to a phenomenon automatically involves ascribing a set of properties to it. To some extent it is hard to avoid some level of reification when speaking and writing about culture or the cultural. Also, and more importantly, reification has more to do with the way concepts are interpreted and applied, than the concepts themselves. The reification of culture is thus not related to the word culture, but to the meaning which people ascribe to the word.

This discussion points towards a constructivist account of culture. According to this view, culture is not something that is *given*; it is *emerging*. This distinction has been largely overlooked by safety culture research. It is illustrated by the fact that few, if any, studies have dealt with the question as to how safety culture is created. Discussions of this question would highlight that cultures are constructed through complex phenomena, and that cultural traits are thus continually recreated and changed.

This constructivist stance should not be misunderstood as implying that nothing is stable, that action and behaviour are performed in a social or cultural vacuum. As authors like Berger and Luckmann (1966) and Giddens (1984) have noted, social structures are both products of social interaction and objectified rules and resources for this interaction.[4] In relation to culture, this implies that culture is socially constructed but that at the same time it will have fairly stable traits, constituting rules and norms for behaviour and decisions. The processes in which cultural aspects are produced, re-created and changed are further described in Chapter 3.

Following these arguments, the concept of safety culture is misleading since there is no such 'thing' as a safety culture. Therefore, strictly speaking, one should abolish the term 'safety culture' (Hale 2000). Rather, we should be studying *organizational* culture and the way it may influence safety, positively or negatively. If one is to use the term safety culture it should be only as a *conceptual label*, referring to the dynamic and complex relationship between culture and safety. This is the way the term is used in this book. The use of the term is thus only for reasons of linguistic simplicity not because it is regarded as referring to some entity or 'part' of the larger organizational culture.

This idea of safety culture, as a conceptual label denoting the positive or negative relationship between organizational culture and safety, implies that I disagree with authors like Hopkins (2005) and Reason (1997), who conceive of safety culture as referring only to those cultures which place high emphasis on safety. In my view, restricting the term to only encompassing a set of positive aspects simplifies the complex relationship between culture and safety and leads directly back to the perpetual discussion of how to define the characteristics of a 'good' safety culture.

I see organizational culture as the primary matter of investigation in safety culture research. This implies a broadening of the scope of safety culture research as it requires a more general analysis of the traits of organizational cultures. Furthermore, the study and description of the cultural traits of a group or an organization comes logically before any discussions of how these traits influence safety.

---

4   This is referred to as 'the duality of structure' (Giddens 1984).

Putting organizational culture at the centre of analysis has the advantage that it connects safety culture research to a much more solid theoretical foundation. This does not mean, however, that the theoretical insights of Turner (1978; Turner and Pidgeon 1997) and Weick (1987, 1999) can be disregarded. On the contrary, the definition of culture given in Chapter 1, by pointing to the frames of reference through which information, symbols and behaviour are interpreted, is heavily influenced by both Turner and Weick. However, the view on culture inherent in the works of Turner and Weick involves a somewhat one-sided emphasis on the cognitive aspects of culture. To be able to grasp fully its social character, a definition of culture must include the normative and relational frames for social interaction. This is why the definition of culture given in Chapter 1 includes an aspect of social conventions for behaviour, interaction and communication. In many ways, this conception of culture represents a synthesis of the way culture is conceived by Turner and Weick, with the practice-oriented conception introduced by Gherardi and colleagues (Gherardi et al. 1998; Gherardi and Nicolini 2000). These writers provide many important insights on how culture may influence safety. They do not, however, deal explicitly with questions of how cultures in organizations are created and re-created. This is why the theoretical tool-box of the safety literature needs to be complemented with contributions from research and theory on organizational culture. In the next chapter I will delineate some important theoretical contributions in this respect.

Chapter 3
Different Perspectives on Organizational Culture

At this point it should have become clear that there is great variety in the way scholars conceive of and study culture. The research on organizational culture is no exception to this. Frost et al. (1991) stress this variety in their summary of the status of research and theory in the field:

> Organizational culture researchers do not agree about what culture is or why it should be studied. They do not study the same phenomena. They do not approach the phenomena they do study from the same theoretical, epistemological or methodological points of view. (Frost et al. 1991: 7)

Many of the differences in theoretical, epistemological and methodological points of view are related to the question of whether culture is something that an organization *has*, or if organizations *are* cultures, analogous to the way nations and tribes can be said to constitute cultures.

Culture as a Variable or Root Metaphor?

In an article from the *Administrative Science Quarterly* special issue on organizational culture,[1] Smircich (1983) discusses the characteristics of these two conceptions of organizational culture. According to her, research which treats culture as something an organization *has* tends to view culture as a distinct *variable*. Similar to structure and technology, culture is seen as a 'component' of organizations. Studies based on this line of reasoning tend to emphasize that culture may have productive functions, especially in terms of efficiency and other aspects of organizational excellence (e.g. Peters and Waterman 1982; Wilkins and Ouchi 1983). This conception of organizational culture is functionalistic as it is oriented to the role culture may play in maintaining functions of internal integration and external adaption by means of providing common identity, commitment and coordination. Smircich sums up her treatment of the culture-as-variable approach by stating that:

---

1   *Administrative Science Quarterly*, 28 (3), September 1983.

> Overall, the research agenda arising from the view that culture is an organizational variable is how to mold and shape internal culture in particular ways and how to change culture, consistent with managerial purposes. (Smircich 1983: 346)

Those who treat organization *as* culture reject this way of seeing culture as being too simplistic, both with regard to the underlying conception of culture and the optimistic view of the possibility for cultural change. It is suggested instead that culture should be seen as a root metaphor for conceptualizing an organization. Smircich describes the essence of this approach as follows:

> Characterized very broadly, the research agenda stemming from this perspective is to explore the phenomenon of organization as subjective experience and to investigate the patterns that make organized action possible. (Smircich 1983: 348)

This view of culture is derived from anthropology, especially from cognitive and symbolic anthropology (Smircich 1983). Its focus is on analysing how members of a culture perceive their world and how they interpret and understand their experience. Organizations are viewed as 'structures of knowledge' (Smircich 1983: 348) or as 'patterns of symbolic discourse' (Smircich 1983: 350). Advocates of this view tend to play down the 'outputs' of culture in terms of some aspects of culture being positive for business. In principle, within this approach, nothing in organizations is 'not culture' (Alvesson 2002). Consequently, it is meaningless to speak of culture having functions for the organization, as culture and organization are two words referring to the same phenomenon.

Smircich's distinction between the two different views on culture points to key differences in the way culture is conceptualized in studies of organizational culture. However, her mapping of the terrain of organizational culture research has some weaknesses. While her overview of the field is highly useful, she misses some important details.

As has been noted by Alvesson (2002), many researchers would fall between the two extremes, refraining from treating organizational culture as a variable, but without fully accepting the view that organizations are cultures. Very few organizational culture researchers today maintain that culture is a variable that can be moulded and shaped by managers, although some of the management literature may have retained some of this optimism. However, this does not mean that there is now general agreement that organizations *are* cultures. There are some fundamental problems associated with viewing culture as metaphor for organizations. This conception of culture runs the risk of reducing everything to symbolism, as it leaves little room for studying the way the organization relates to its market situation, its material conditions and its external environment (Alvesson 2002). This has to do with where one is to draw the line between cultural and non-cultural phenomena. If there are no phenomena that are not cultural, then the concept of organizational culture is certainly void of any meaning, since 'culture' and 'organization' would be synonymous terms.

In my view, it is fruitful to reserve the concept of organizational culture for informal aspects of an organization, thus distinguishing between organizational culture and elements of organizational structure and design. Although organizational structures are certainly human products, they are not influenced by the day-to-day interaction of 'ordinary' members of the organization. This means that the organizational structures are only loosely connected to much of the culture-producing activity of the organization. Organizational structures and design are best conceived of as created in the intersection between the cultural processes of managers and the dominant beliefs of the organization's institutional environment (Meyer and Rowan 1977). Structures are thus influenced by cultural processes in that their creation is determined by the frames of reference and conventions for behaviour and decision making of those producing them. Nevertheless, structures are different from aspects of culture in that they take on a formalized status in the organization. I therefore reserve the term culture to refer to informal aspects of work and organizing. This does not mean that I think formal structures and organizational design are irrelevant for the study of organizational culture. On the contrary, and as I will elaborate further in the remainder of the book, I see the relationship between formal and informal organization as key to the study of organizational culture.

Whether culture is to be seen as a variable or a root metaphor is not the only topic debated within organizational culture theory and research. Another line of demarcation is drawn around the question as to whether organizational culture should be studied through the lens of integration or conflict. And, as well, there are researchers emphasizing a more postmodern approach, claiming that organizational cultures have become increasingly transient and that issues of ambivalence and ambiguity should be included in the study of organizational culture (Martin and Meyerson 1988). These three ways of viewing organizational culture are labelled the 'integration perspective', the 'differentiation perspective', and the 'fragmentation perspective' (Martin and Meyerson 1988; Frost et al. 1991; Martin 1992, 2002). The division of the field into these three perspectives provides a useful frame for describing the empirical research on organizational culture. I shall delineate the key characteristics of each of these perspectives, starting with the integration perspective.

The Integration Perspective

Research from an integration perspective on organizational culture often views culture as the social 'glue' of the organization, i.e. a precondition for cohesion, or as a 'compass' which gives direction to and coordinates the actions of the organizational members (Alvesson 2002). Schein is the most well-known proponent of the integration perspective. He defines culture as:

A pattern of shared basic assumptions that the group learned as it solved its problems of external adaptation and internal integration, that has worked well enough to be considered valid and, therefore, to be taught to new members as the correct way to perceive, think and feel in relation to those problems. (Schein 1992: 12)

According to Frost et al. (1991: 13), such integration-oriented definitions imply that culture is characterized by 'clear and consistent values, interpretations, and/or assumptions that are shared on an organization wide basis'. In other words, the integration perspective emphasizes the existence of a high degree of *consensus* between the members of the organization; internal *consistency* between different cultural aspects and cultural traits are assumed to have a *clarity* that makes it possible for them to be understood in the same way by different organizational members (Martin and Meyerson 1988; Martin 1992). In order to delineate the contents of the integration perspective, the criteria of consensus, consistency and clarity need to be examined in some detail.

Consensus   According to Martin (1992), the integration perspective's emphasis on consensus involves an assumption that there exists a set of values and understandings on which there is organization-wide agreement. This agreement is assumed to span across the different departments and hierarchical levels of the organization. This consensus is perceived to be necessary to reduce friction and conflicts between the members of the organization. At this point it must be stressed that there is great variety in how researchers perceive the level of consensus in an organization. Studies from the early 1980s (e.g. Deal and Kennedy 1982; Peters and Waterman 1982) describe organizational culture as a near-total consensus regarding the organization's espoused values and visions. In more recent integration-oriented studies (e.g. Schein 1992, 2003), the level of consensus assumed is far more moderate. This issue will be elaborated further in the discussion of the relationship between the three perspectives.

Consistency   Those who adhere to an integration perspective see the relationship between different aspects and expressions of culture as characterized by internal consistency. This pertains not only to the relationship between different rituals, symbols and values, but also in the relationship between culture and thinking, perception and behaviour. Schein's (1992) model of culture as consisting of different layers is an example of this feature. The visible layers of culture, what Schein calls artefacts such as language, rituals, dress codes, architecture and technology, are seen as expressions of the organization's values, e.g. strategy, business philosophy and goals, which in turn rest on a set of basic assumptions, taken-for-granted worldviews, that constitute the cultural core (see Figure 5.1 for a similar model). Although Schein's writings sometimes indicate some degree of conflict between artefacts, espoused values and basic assumptions, see his criticism of research on organizational climate delineated in Chapter 2, the underlying assumption of the model is that the superficial levels can only be understood through investigating the underlying layers. According to Schein's model once all the layers of organizational culture are investigated, their internal coherence will become visible. This consistency implies that the different elements of an organization's culture will cover the same themes and will mutually reinforce one another.

Clarity   As I mentioned previously, a compass is one of the metaphors used to describe what culture is and the way it relates to thinking, perception and behaviour. For culture to have such a function, the members of the organization must be able to perceive and interpret cultural expressions in more or less the same way. In other words, according to the integration perspective on culture, aspects of culture must display a high degree of clarity. It is on the basis of the compass metaphor that some have compared the integration view on culture to a monolith (see Richter and Koch 2004). A monolith is a pillar of rock, chiselled out of a single block, that has the same shape irrespective of from which side it is viewed. According to the integration perspective, or perhaps more correctly according to the critics of this perspective, an integration view of culture assumes that traits of culture will be unitary and 'look the same' irrespective of where the 'viewer' is positioned in the organization. This view on culture is often seen in relation to basic human needs for meaning and predictability. From an integrative viewpoint, culture fills an important function in this respect, as it serves to 'alleviate anxiety, to control the uncontrollable, to bring certainty to the uncertain and to clarify the ambiguous' (Martin 1992: 51). This quotation illustrates that research within an integration perspective tends to exclude ambiguity and ambivalence from the definition of culture. The exclusion of ambiguity and ambivalence from the *definition* of culture does not imply that it cannot be included in the *study* of culture as I shall discuss below.

Culture is seen as a mechanism that *reduces* ambiguity and ambivalence in organizations and all other aspects of social life. Consequently, a high degree of ambiguity and ambivalence will imply that there is an absence of culture. Whether it is meaningful to conceive of culture as ambiguous has been subject to much theoretical discussion and constitutes the fundamental line of demarcation between the integration and the fragmentation perspectives on organizational culture.

Integration, leadership and organizational performance   Some, but far from all, integration-oriented studies tend to be somewhat lopsided in emphasizing the assumed positive functions of culture. Studies by Wilkins and Ouchi (1983), Deal and Kennedy (1992) and Peters and Waterman (1982) are widely known examples from this literature. Many would also include the writings of Schein in this category, although to be fair, Schein bases his writings on a far more advanced conception of culture than the popular literature on corporate culture. The essence in these studies is that a highly integrated organizational culture is viewed as functional for the organization's ability to produce high-quality products and services, communicate effectively, be innovative and for other aspects of organizational performance (Alvesson 2002). It is assumed that organizations characterized by cultural integration will be easier to manage and coordinate and have a lower level of conflict and a higher level of commitment to the organization's goals than those with less integrated cultures.

The idea that a strongly integrated culture will make organizations more efficient has also led some adherents of the integration perspective to move from being descriptive to being *prescriptive*. Some of the popular literature from the 1980s presents 'recipes' for how organizational culture may be used as a management tool (e.g. Deal and Kennedy 1982). The logic of this line of reasoning is that the right results can be achieved by building the 'right' culture. Even though some integration-oriented research still emphasizes that leaders usually play important roles in the creation of culture (e.g. Schein 1992), very few current researchers on organizational culture see organizational cultures as something that can be changed according to managers' discretion. For instance, the early literature on corporate culture has been criticized for being manipulative and therefore ethically problematic. The historian Francis Sejersted, for instance, has noted that Peters and Waterman's bestseller *In Search of Excellence* (1982) in reality recommends managers '[a]t the same time [to] brainwash their employees *and* treat them as individuals. The possible antagonism between these two strategies is not discussed.'[2], [3] (Sejersted 1993: 2, author's emphasis).

This line of criticism introduces the notion of power and conflict to the study of culture. Critics of the integration perspective have accused it of building on a harmony model of organizational life that does not correspond to reality. They argue convincingly that most organizations are *not* characterized by harmony and community but are arenas for conflict and power struggles between groups with contending interests. From a conflict-oriented standpoint, the existence of organization-wide cultural consensus is highly unlikely. An organization's culture will, according to this perspective, be differentiated into several subgroups that may stand in opposition to one other. Probably inspired by writings that view organizations as political systems (e.g. Cyert and March 1963), research within such a differentiation perspective aims to show that organizational life is characterized by conflicts of interests and power dynamics and that these aspects should be included in the study of organizational culture. The way the concept of power can be incorporated into the study of safety culture is discussed in Chapter 4.

The Differentiation Perspective

While the concept of organizational culture within the integration perspective is usually taken to refer to one, organization-wide, culture, research conducted

---

2   My translation.

3   In addition to being a renowned historian, Sejersted was also a leading ideologist within the Norwegian conservative party. Coming from an ideologist of the political right adds further weight to the somewhat radical message in the quotation.

within a differentiation perspective views organizations as consisting of multiple *subcultures*. Van Maanen and Barley (1985) define a subculture as:

> ... a subset of an organization's members who interact regularly with one another, identify themselves as a distinct group within the organization, share a set of problems commonly defined to be the problems of all, and routinely take action on the basis of collective understandings unique to the group. (Van Maanen and Barley 1985: 38)

Van Maanen and Barley's definition illustrates that the concept of culture inherent in the differentiation perspective is inextricably connected to the notion of social interaction, an aspect less emphasized in the integration perspective. In organizations over a certain size, the complex division of labour will make it very difficult to coordinate activity by means of mutual adjustment. Rather the design of larger organizations will most often involve different forms of divisionalization and hierarchical models of organizing (Mintzberg 1983). This means that interaction and communication will take place in subgroups within the larger organization. In turn, this makes possible the creation of several local cultures within the organization. These may also be connected to communities which cut across the boundaries of singular organizations, such as occupational cultures. In this respect, organizations are perhaps better seen as the nexus of different cultures, national, regional, occupational etc., rather than being viewed as cultural units themselves (Martin 2002). This view of organizations as consisting of different cultural camps involves a very different answer to the questions of consensus, consistency and clarity compared with that provided by the integration perspective.

Local consensus Since studies within a differentiation perspective emphasize conflicts of interests and differences of power, they obviously take issue with the harmony model of organizational life characteristic of the integration perspective. Even though the *possibility* of organization-wide consensus is not explicitly rejected, a differentiation approach will view organizations as characterized by dissent rather than consensus. Within this perspective, consensus will primarily be something *local*, within the boundaries of a subculture (Gregory 1983; Martin and Siehl 1983; Frost et al. 1991; Martin 1992). Differentiation studies typically focus on consensus within groups located at the bottom of organizational hierarchies (cf. Lysgaard 2001). In many ways, differentiation studies represent a critical, almost emancipatory, approach to the study of organizational cultures since they aim to uncover conflicts and power relationships that may reside in seemingly harmonious organizational contexts. Van Maanen's study *The Smile Factory* (1991) shows how the idyllic façade of Disneyland covered subcultural dynamics characterized by far from harmonious relationships. Within safety research, a similar, critical approach has been advanced by Perrow's (1999) emphasis on the role of power relations in the generation of accidents.[4] Power relations are probably especially important in the organizations studied in this book; the Norwegian petroleum industry has a long history of strong unions that may not always agree with managers' perceptions of risk and safety (e.g. Ryggvik 2003). Although in Norway there is a long tradition of industrial democracy and collaboration at work between unions and employer organizations, the power relations between unions and management in the petroleum industry represent a form of institutionalized differentiation. The dynamics of power struggle should therefore not be underestimated.

This critical dimension is probably the reason why Van Maanen and Barley (1985) go a long way in questioning the utility of the concept of organizational culture. They argue that since the concept is often used to refer to organization-wide unity and integration at the same time as the unitary concept of culture finds little empirical support, the concept should be abandoned altogether. While I certainly agree with the premises of this argument, I do not agree with the conclusion. Van Maanen and Barley's criticism is, in my view, more striking with regard to the way the concept is used by some authors than the concept in itself. If one were to abandon every concept that was improperly or imprecisely used I suspect that social science would quickly run out of concepts. Since our theoretical toolbox is formulated in the same language as everyday talk and writing and not some objective or mathematical language, the misuse of concepts is unavoidable.

Inconsistency   Where integration-oriented studies view cultural aspects as integral parts of a functional whole, differentiation studies focus on the instances of inconsistency between different aspects of an organization's culture. Martin (1992) emphasizes three forms of inconsistency that characterize the differentiation perspective on culture. First, in most organizations it will be possible to identify inconsistencies between words and action, e.g. between the official values and real-life practice. This difference between theory and practice corresponds to the discrepancy between espoused theory and theory-in-use described by Argyris and Schön (1996). It can also be seen in relationship to Goffman's (1959) concepts of the 'frontstage' and 'backstage' of social life.

Second, instances of *symbolic inconsistency* are of key importance within the differentiation perspective (Martin 1992). All cultural expressions are subject to interpretation, and may be interpreted differently by different members of a culture. Organizational myths and narratives may convey radically different messages depending on who relates them and who listens. In the context of safety, stories that, for workers, symbolize toughness and courage may be seen as examples of stupidity and arrogance by safety managers. The third form of inconsistency described by Martin is *ideological inconsistency*, which refers to conflicts between two or more of an organization's values or basic assumptions. This form of inconsistency is particularly important for the study of the relationship between

---

4   See in particular his discussion on the utility of the concept of safety culture in the afterword to the 1999 edition of *Normal Accidents* (Perrow 1999: pp 378–380).

organizational culture and safety, since the tension between the prioritization of safety and cost efficiency can be seen as an instance of such inconsistency. The tension between goals of safety and goals of profit has proven to be an important factor in most major accidents, for example Medvedev's account of the Chernobyl accident (1991).

Clarity only within subcultural borders   A third characteristic of the differentiation perspective lies in its account of ambiguity (Martin and Meyerson 1988; Frost et al. 1991). When subcultures are characterized by internal consensus, this implies that there will be a high degree of clarity in the aspects of local culture. Thus, the differentiation perspective, like the integration perspective, views ambiguity as an absence of culture. Ambiguity is thus excluded from the definition of culture; it is seen as an aspect primarily having to do with the external environments of subcultures.

Martin (1992) sums up the characteristics of the differentiation perspectives in the following way:

> To summarize, Differentiation research describes each subculture as an island of localized lucidity. ... In this way, subcultural differentiation 'fences in' differences in perspective, leaving uncontrolled, untransformed ambiguity 'outside' or 'underneath' these realms of clarity, in the interstices between subcultures. (Martin 1992: 93)

That the integration and differentiation perspectives exclude ambiguity from the definition of culture is the main line of criticism of the so-called fragmentation perspective. The main argument of fragmentation-oriented researchers is that instead of viewing cultures as 'islands of localized lucidity' (ibid.), cultures may in fact be characterized by a lack of such lucidity. Consequently, issues of ambiguity and ambivalence should be regarded as key issues of the study of culture.

The Fragmentation Perspective

The fragmentation perspective is the most recent and also the most radical of the three perspectives on organizational culture. Researchers within this perspective (e.g. Feldman 1991; Meyerson 1991) have a fundamentally different view on the degree of consensus, consistency and clarity inherent in the concept of culture compared with the two previous perspectives. Although it is difficult to define culture within a fragmentation perspective, Martin (1992) has attempted to give a formal definition:

> From a Fragmentation perspective, then, an organizational culture is a web of individuals, sporadically and loosely connected by their changing positions on a variety of issues. Their involvement, their subcultural identities, and their

> individual self-definitions fluctuate, depending on which issues are activated at
> a given moment. (Martin 1992: 153)

Fragmentation-oriented studies criticize the integration and differentiation perspectives for systematically excluding ambiguity, inconsistency and ambivalence from their analyses of culture. They claim that those two perspectives both overemphasize consensus, either in the organization as a whole or within subcultural boundaries. In contrast, from a fragmentation perspective culture is viewed as polyphony of interpretations which rarely coincide to form a stable consensus (Frost et al. 1991).

The analytical focus of fragmentation studies is on describing the multiplicity of interpretations that integration studies rarely address and that differentiation studies reduce to an aspect of the dynamic between subcultures. This line of reasoning addresses some important features of post-modern organizations. The boundaries of organizations tend to become increasingly blurred through the use of outsourcing strategies and temporary employment. This has given some of them the characteristics of being more like loose networks than stable hierarchies (Castells 2000). At the same time there has been a growth in the number of so-called distributed organizations, i.e. organizations where the employees are no longer geographically co-localized and only communicate by means of ICT (Hinds and Kiesler 2002). Some claim that we now see a new form of working life, characterized by swift and frequent changes (Dale-Olsen 2005). Taken together, these new forms of organizing constitute a different organizational reality compared with traditional organizations. Martin (1992) sums up these consequences in the following way:

> Taken together, these factors create an organizational world characterized by
> distance rather than closeness, obscurity rather than clarity, disorder rather than
> order, uncontrollability rather than predictability. (Martin 1992: 132)

According to fragmentation theorists, these radical societal and organizational changes imply that culture must be studied through correspondingly radical analytical perspectives and theories. Organizations which have blurry borders and are in continuous and rapid change cannot be studied through analytical lenses which assume that cultural patterns are stable and that organizational members make concurrent interpretations of the world surrounding them.

Since ambiguity is the key concept of the fragmentation perspective, that appears as a fuzzier framework than the other two, it is necessary to clarify what is meant by it in this perspective on organizational culture.

Ambiguity Ambiguity arises when there are no clear interpretations of a phenomenon or course of events (Feldman 1991). Ambiguity thus has to do with a lack of consensus, consistency and clarity. March distinguishes between ambiguity in situations, purposes, identities, results and histories. He describes the various types of ambiguity in the following way:

> Ambiguous situations are situations that cannot be coded precisely into mutually
> exhaustive and exclusive categories. Ambiguous purposes are intentions that
> cannot be specified clearly. Ambiguous identities are identities whose rules or
> occasions for application are imprecise or contradictory. Ambiguous outcomes
> are outcomes whose characters or implications are fuzzy. Ambiguous histories
> are histories that do not provide unique, comprehensible interpretations. (March
> 1994: 178)

As has been noted by Weick (1995), the common denominator for different
forms of ambiguity is that they involve a suspension of the conditions for rational
decision making. Faced with ambiguity, actors are left without clear guidelines for
determining which action constitutes the best alternative. All the described forms
of ambiguity may arise in organizations. Pertaining to organizational culture,
ambiguity implies that organizations' histories, identities, symbols, values and
basic assumptions can, and will, be interpreted in an infinite number of ways.

In ambiguous situations, every actor in processes of sensemaking is left alone
in constructing his or her own definition of reality (Weick 1995). The process of
making sense of ambiguous and complex situations is described by Weick as:

> ... a constant alternation between particulars and explanations with each cycle
> giving added form and substance to the other. ... The image here is one of people
> making do with whatever they have, comparing notes, often imitating one
> another directly or indirectly. (Weick 1995: 133)

Within such a perspective on organizational life there is no predefined cultural
script providing guidance for behaviour. The fragmentation perspective thus
emphasizes the situational and unpredictable aspects of action and interaction. In
this respect, the fragmentation perspective differs radically from the integration
and differentiation perspectives that place far more emphasis on the structures
providing guidance for social action and interaction. This illustrates that the
differences between the three perspectives touch upon the classical lines of
division in sociology; those running between actor-oriented and structure-oriented
perspectives on social action, between constructivism and functionalism and
between integration and conflict.

The Relationship Between the Perspectives

It should be clear from the above account that the integration, differentiation and
fragmentation perspectives on organizational culture give very different accounts
of what culture is and its role in organizations. The issues of conflict and ambiguity
are among the key differences between the three perspectives. These differences
represent more than just variation in research interests, as the three perspectives

build on different epistemological and ontological assumptions about the 'nature' of culture, organizations and society. This is bad news for those wanting an overarching theoretical perspective welding the research on organizational culture into one theoretical frame.

In their 'pure' form, the three perspectives are analytically incompatible in key respects. This does not mean, however, that elements from the various perspectives cannot be combined. After all, the three perspectives share a common interest in the expressive and non-rational aspects of work and organizing (Smircich 1983). The three perspectives can also be said to capture different aspects of the same subject matter, i.e. they are abstractions of the same phenomenon. With this line of reasoning as a starting point, Frost and colleagues have argued that the three perspectives can be employed as complementary perspectives in a process of theoretical *reframing*. Their argument is that combining the three perspectives can provide insights that are not possible to 'see' by means of one perspective alone (Frost et al. 1991). I fully agree with this argument. However, as I will discuss in the next sections, I do not share their view that this requires a process of reframing.

Social Integration versus Conflict

The criticism by the theorists for differentiation that the integration perspective systematically ignores conflicts in the organizations studied (Meyerson and Martin 1987; Martin 2002), is certainly justified with regard to the popular literature on organizational culture of the 1980s. But there is, however, a lot of serious research within this perspective, something which is not always credited when critics such as Richter and Koch (2004) describe the integration perspective. Moreover, this research, while being based on an integrative conception of culture, does not preclude the issue of conflict. Schein, for instance, is careful to include some elements of both conflict and fragmentation in his analyses of organizational culture. This is evident from his discussion of the question as to whether large organizations can be expected to have one culture: 'Our experience with large organizations tells us that at a certain size, the variations among the subgroups are substantial, suggesting that it is not appropriate to talk of "the culture"' (Schein 1992: 14).

Schein here also adds that there is no clash between the existence of shared basic assumptions and subcultural conflicts. This is an important clarification, often overlooked by differentiation theorists. Even though integration-oriented research admittedly does not place the issue of conflict at the centre of analysis, this does not imply that its very existence is denied. Thus, the differences between the integration and differentiation perspectives have to do with where one draws the line as to what constitutes a cultural unit. The underlying conception of culture is basically the same.

There is, nevertheless, a very real danger that purely integrative studies might paint unrealistically harmonious pictures of organizations. The expression 'seek, and thou shalt find' certainly applies to studies of organizational culture. If one is looking to investigate *only* the aspects that are shared among all employees, then one will most likely overlook the overt or latent lines of conflict that are important to investigate in order to understand how the organization works. The point being made here is that the existence of conflict in some areas does not preclude the existence of something shared in others. To understand the cultural dynamics in organizations, one will need an analytical approach including both aspects of integration and aspects of conflict and power.

Fragmented Cultures or Fragmented Organizations?

Studies of fragmentation often describe situations or action sequences where individual actors are faced with ambiguous stimuli. Weick's (1991) study of the accident at Tenerife airport where two passenger aircraft collided on the runway is one example. In this study, Weick showed how a stressful work situation created a high degree of complexity and linguistic confusion between the air traffic controllers and the flight deck crews of the two planes. This created a breakdown in coordination leading to the two planes colliding and the deaths of 583 people. It is a striking feature of this study that the ambiguity and complexity described by Weick are properties of the situation and activities involved. The situational complexity and ambiguity were caused by the planes being hastily redirected from the international airport of Las Palmas to the much smaller airport of Tenerife. This airport turned out to be too small to deal with the large amount of traffic redirected from Las Palmas. The situation grew even more complex when very dense fog came in over the runway. The point here is that the ambiguity created by these coinciding factors is an attribute of the situational context, *not* cultural aspects of the organizations involved. It is indicative in this respect that the word 'culture' is not mentioned once in this study, despite being published in a book about organizational culture.[5]

Weick's study highlights an important question regarding the fragmentation perspective: what is really fragmented – the culture or the organization? This is a key issue in the relationship between the fragmentation and the integration and differentiation perspectives. Fragmentation theorists would regard this to be something of a 'trick question'; they view organizations as *being* cultures and they see no distinction between the two terms. In my view this is not the case, as it would involve taking too literally the standpoint of culture as a root metaphor of organizations. By equating organizations that have no influence on the primary socialization of their members with national or tribal cultures where primary socialization takes place within the boundaries of the culture, a crucial aspect is lost. In such a conception, culture is no longer treated as a metaphor for organizations; it is treated as *synonymous* with organizations.

Against this line of criticism, Martin (1992: 132) defended the fragmentation perspective by stating: 'If theory and research are to be relevant to problems of

---

5   The study is to be found in Frost et al. (1991).

contemporary organizational life, the exclusion of ambiguity cannot be an option'. This is indeed a valid argument, as contemporary organizations are no doubt affected by the swift changes in society and working life. Fragmentation theorists like Martin are also correct when they argue that ambiguity and the polyphony of interpretations must be included in the study of organizational culture, and that organizational members are not mere passive recipients of culture. Indeed, the integration perspective tends to view culture as an already existing framework for action, illustrated by Schein's definition of culture as a set of 'basic assumptions that the group learned as it solved its problems of external adaption and internal integration' (Schein 1992: 12). It is indicative of Schein's somewhat static conception of culture that he consequently writes on the creation of culture only in the past tense.

I see the fragmentation perspective as an overreaction to the integrative approach to organizational culture. Fragmentation theorists go too far in emphasizing what is transient at the expense of what is stable. One might ask what is left of the concept of culture if it is to be defined by pointing to that which is *not* shared among the members of a group. A concept of culture which does not include any form of common foundational schema (Shore 1996; Almklov 2005) is in my view a concept void of meaning. Goodenough has made a similar point with regard to the degree of 'sharedness' which must be present if it is to be meaningful to see it as a cultural phenomenon:

> People who interact with one another regularly in a given kind of activity need to share sufficient understanding of how to do it and communicate with one another in doing it so that they can work together to their satisfaction. In fact, all they need to share, is whatever will enable them to do that. (Goodenough 1994: 266)

Note that the emphasis on culture as something that is by definition shared pertains only to the *definition* of culture. It does not mean that the *study* of culture can disregard issues of ambiguity and fragmentation. On the contrary, it is essential to contemplate these issues in order to investigate the degree to which something is or is not shared among the members of a group or an organization and the degree of clarity of and consistency between cultural expressions.

On the Need for Analytical Reframing

The distinction between the three perspectives provides a highly useful means of classifying different studies of organizational cultures according to which of the perspectives are given prominence in them.

While fully agreeing on the need to include the insights of all three perspectives in the study of organizational culture, I do not, as I said earlier in this chapter, see the need for the reframing approach suggested by Frost et al. (1991). As the integration and differentiation perspectives share an underlying conception of culture as something that the members of a group share, the differences between these two perspectives are not so substantial as to require that they be kept apart. As I have already indicated, the shift between analysing the cultural traits of a whole organization and those of a subculture is more to be regarded as a shift in the cultural units studied than a shift of analytical perspective.

The inclusion of ambiguity into the *study* of culture does not require a process of reframing either, as long as ambiguity is not included in the *definition* of culture. The inclusion of ambiguity in the study of culture is highly necessary to determine what is shared and what is not. If there are no shared meanings in an organization then the organization has no culture. Thus, in principle, the presence of culture in an organization is an empirical question but an organization must probably have some degree of shared culture if any organizing processes are to be possible (Alvesson 2002).

My treatment of ambiguity is inspired by Alvesson's notion of 'bounded ambiguity', which refers to cultures in which there is not necessarily any established form of consensus, consistency and clarity among broad groups of people, but where there still exist some cultural guidelines for coping with instances of ambiguity and ambivalence. Within this perspective, while experiences of ambiguity are not excluded from the study of culture, although they are from the definition of it, it is still assumed that the existence of culture presupposes some form of 'shared meanings and efforts to minimize the most disturbing experiences of confusion, contradiction and notorious uncertainty' (Alvesson 2002: 166). If nothing else, the members of such cultures share some orientation on how to deal with uncertainty and ambiguity.

So What is Organizational Culture?

At this point in discussing the various approaches to the study of organizational culture it is time to return to the key question of the 'nature' of culture, in order to further clarify my position in relation to the lines of division I have set out here.

With regard to the distinction between organizational culture as a variable and culture as a root metaphor for organizations, I take issue with both positions. The culture-as-variable position is untenable because culture is not something that can be moulded and manipulated by managers, and if it were, in turn, it would mean reducing complex social phenomena to an entity defined by a set of fixed properties. Such reification would involve a 'fallacy of misplaced concreteness' (Whitehead 1929, in Almklov 2005), i.e. treating ideational phenomena as if they were concrete objects. Viewing culture as a root metaphor of organizations avoids this fallacy but the approach goes too far in the other direction. There are clearly many characteristics that separate organizations from what within anthropology were originally seen as cultural units, for instance the Masai people or the Yir Yoront tribe. This is particularly the case since organizational cultures will always be rooted in a larger regional and national culture and will not display the same degree of distinctiveness and uniqueness as national or tribal cultures. In my view these differences imply that organizational entities and nations or tribes cannot be seen as belonging to the same class or level of cultural phenomena. This indicates that the metaphor of organizations *as* cultures should not be exaggerated. This does not mean, of course, that organizations are not settings for cultural processes nor that they consequently cannot develop distinct cultural traits. All human activity produces cultural results and is influenced by cultural frames.

While culture is certainly an *aspect* of organizations, it cannot be reduced to a subsystem of organizations. There are nevertheless important differences between instances of organizational culture and national/tribal culture. These lie in how deeply the cultural frames are rooted in the individual and the degree to which they are taken for granted by the members of the cultural unit.

How are Cultures Created and Changed?   I have not said much so far about the processes through which culture is created apart from indicating a constructivist stance. Taking a constructivist perspective means that I see culture as created through the day-to-day interaction between the members of a given community. The process through which social interaction creates lasting patterns for behaviour has been described by Berger and Luckmann in their famous work, *The Social Construction of Reality* (1966). According to them, social reality is both a subjective and objective phenomenon. It is subjective in the sense that it is the product of human behaviour – the social structure and culture around us are 'webs of significance' that we ourselves have spun, to use the words of Geertz (1973). At the same time, social reality is objective in the sense that it existed in some form prior to each individual member of society or an organization and that these structures of meaning are reified by the individual actors, i.e. they are perceived as if they were *not* human products.

Social reality appears, in other words, as objective to the members of the community in question, but is nevertheless a product of human interaction at the same time as the individuals are to be regarded as products of society (Kalleberg 1985). The main steps in the process of the social construction of reality are described in Figure 3.1.

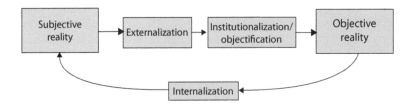

Figure 3.1   Berger and Luckmann's (1966) perspective on the social construction of reality

I would describe the three stages of this model as follows: Externalization represents the way we project our interpretation of the world into the world through our acts and statements. By this process we make our interpretation of the world public to other actors who may, in turn, build some expectation about how yet others will act or behave in the future. These expectations form the basis of habitualized action, i.e. recurrent patterns of action and interaction. In time, these patterns of habitualized action may become invested with meaning and become a way of doing things that stands out as *right* for the actors involved. They may even be considered as the *only* ways to do things. This is the process of institutionalization. This is where the social patterns take on a life of their own. The institutionalization process makes the patterns become a part of the objective reality *around them*, they become objectified. Through processes of primary and secondary socialization, the objectivated social world is retrojected into the individuals' consciousness. This feedback loop is labelled internalization by Berger and Luckmann (1966). The dialectic relationship between the processes of externalization, objectification and internalization illustrate that this model of the genesis of social reality is not static; each round of this cycle will involve active interpretation and changes in the existing patterns.

Berger and Luckmann's model aims to explain the existence of a stable and orderly society. This is a somewhat wider enterprise than explaining the genesis of culture. However, the very essence of their thinking is that all human action and interaction has important spin-off effects in that it creates some expectations or 'rules of the game' for future interaction. This is the common starting point for several key sociological concepts, such as those of social roles, norms, stratification and, also, culture. Thus, I view Berger and Luckmann's model of the social construction of reality as a valid explanation for how cultures are created, re-created and changed.

One addition needs to be made to Berger and Luckmann's model, and that concerns the role of power. There has been some debate among social anthropologists as to whether constructivist perspectives may be taken to imply a democratic view on the creation of culture; that all actors contribute equally to its creation. Against Geertz's conception of culture as socially constructed 'webs of significance', Keesing (1987, 1994) argued that factors like power and ideology tend to be ignored. This argument represents a Marxist view, emphasizing that the production of culture is by no means unaffected by social structure. This is a very important line of criticism, echoing some of the arguments of the differentiation perspective outlined earlier in this chapter. In addition to the differentiation theorists' emphasis on the potential conflicts *between* cultures, Keesing's argument introduces another dimension, i.e. that power and conflict influence the very content of culture. The relationship between culture and power with regard to safety will be further analysed and discussed in Chapter 4.

This view of the production of culture implies that culture will, by definition, be unruly and unpredictable. As it is produced, reproduced and changed through everyday interaction *not* by strategic decision-making, culture cannot easily be modified by managers. It is possible, however, to change the 'growing conditions' of culture, which in turn may lead to cultural change, but the results of such changes will forever be unpredictable as there will always be a number of factors influencing the growing conditions of culture.

The relationship between culture and other aspects of organizations   Since I do not fully share the view that organizations *are* cultures, there will be some aspects of organizations that are not cultural and some remarks on what constitutes 'the rest' are needed. As a gross simplification, one might say that organizations have three different 'components': culture, structure and interaction. This categorization corresponds to the 'triangle of social reality' introduced by Boudreau and Newman (1993), depicted in Figure 3.2. These three generic aspects of organizations constitute the organizational framework for all organizational life.

The properties of culture have been duly discussed above and need not be repeated here. Mintzberg defined organizational structure as follows:

> The structure of an organization can be defined simply as the sum total of the ways in which its labor is divided into distinct tasks, and then its coordination is achieved among these tasks. (Mintzberg 1983: 2)

In other words, organizational structures specify the vertical and horizontal distribution of tasks, roles, responsibility and authority. In addition to these dimensions of structure, I would also place under this heading more physical structures like technology.

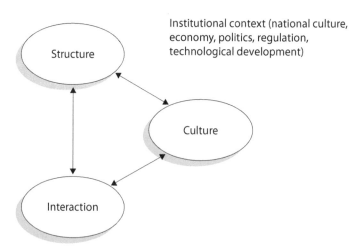

Institutional context (national culture, economy, politics, regulation, technological development)

Figure 3.2   Overview of aspects of organizations

Adapted from Boudreau and Newman (1993).

Interaction is the process through which social relationships and roles are created and acted out, and socialization takes place (Schiefloe 2003). Communication and the exchange of information are also key parts of interaction (Boudreau and Newman 1993). Interaction may be characterized by properties like cooperation, conflict, competition or trust. As I have already indicated, I see social interaction as the key process in the construction of culture.

The close relationship between interaction and culture indicates that dividing the properties of organizations into such categorizations is only an analytical distinction. In real life, the different aspects of organizations are inextricably intertwined. Importantly, issues of power pervade every aspect of an organization as well as its environment.

Organizational culture and safety How do we know that organizational culture influences safety? This is, first of all, an empirical question as we do not know anything about this relationship in actual organizations until its possibility has been investigated. Several accident investigations have delineated several cultural characteristics which may influence safety negatively. Also, Turner's (1978) grounded theory of the cultural contribution to the creation of accidents, as well as the literature on HROs, contain important insights as to how culture may influence safety.

Organizational culture cannot, however, be studied in isolation from the structural and interactional aspects of organizations. To include the interplay between the different aspects of organizations is particularly important with regard to the study of safety. Safety is a composite phenomenon and it will be impossible to study the cultural effects in isolation from the structural and interactional effects. The study of safety culture must thus also include the interplay between these aspects. On this basis it is possible to construct a simple model for the study of safety culture, shown in Figure 3.3.

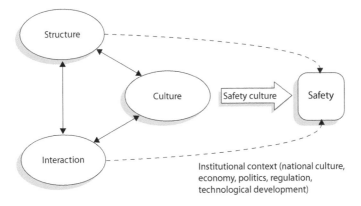

Figure 3.3   Overview of aspects or organizations and their relationship to safety

Remember that the term safety culture is used only as a conceptual label to denote the relationship between organizational culture and safety. Note also that structural and interactional aspects of organizations may influence safety without culture intermediating, expressed through the dotted lines. These possible safety effects lie, however, outside the scope of this book.

This model is, of course, very general and a simplification of the complex and multifaceted nature of organizational safety. In particular, the institutional context of organizations is 'black-boxed' in this model. Organizational cultures do not exist in a social vacuum. The cultural traits and processes that can be found within organizations are influenced by, and also influence, aspects of national culture and the risk governance regimes of regions, countries and industries. The reader should bear this in mind throughout the remainder of the book. I have chosen to focus mainly on processes that take place within organizations since the aim of this book is to shed light on the relationship between culture and intra-organizational processes. So as to avoid merely scratching the surface on this issue, institutional aspects are kept somewhat in the background.

The theoretical platform described in this chapter pertains to organizational culture in general. The link to safety has yet to be demonstrated. The next chapter takes the first steps in doing this.

Chapter 4
Safety Culture and Power[1]

Chapter 3 showed that issues of differentiation in general and power in particular lie at the heart of the theoretical discussion regarding organizational culture. The same cannot be said, unfortunately, of theorizing and research about safety culture. The aim of this chapter is therefore to investigate the relationship between cultural and power-oriented approaches to organizational safety and to make a case for incorporating the issue of power into the study of safety cultures. The argument is that without taking into account the issue of power, safety culture research is in danger of creating superficial and simplified representations of organizational life. This would imply falling into the same pitfalls as the mother concept of organizational culture did in the early 1980s, where managerialist 'quick fix' approaches threatened both the practical and scientific value of the concept.

As I have already indicated, few studies of safety culture include the notion of power within their analytical scope. This does not imply, however, that there has been no emphasis on issues of power in relation to safety. For instance, Sagan (1993) mentions conflicts of interests as one of the key limits to safety. But with the exception of the polemics between researchers of High Reliability Organizations (c.f. LaPorte and Consolini 1991) and Normal Accident Theory (Perrow 1984), there has been little communication between the power-oriented and culture-oriented lines of research. Indeed, some power-oriented theorists seem to question the relevance of culture to understanding organizational safety. Charles Perrow, one of the grand old men of organizational safety research, was once asked why he did not mention culture in his well-known book, *Normal Accidents* (1984). He replied:

> I have not written anything explicitly on the culture because I doubt its utility. It wasn't 'shared values and beliefs' that overruled safety engineers at Ford and Firestone about dangerous tires; it was top managements(sic) concern for profits that hid the data from US government and lied about it ... Of course there are 'cultures' (*note the plural*) in companies, but on issues of risk and safety I think the issue is really power.

He elaborates this view in the afterword to the second edition of *Normal Accidents*, where he comments on the recent growth of interest in the role of culture, by

---

1   This chapter is partly based on an article published in *Safety Science* – see Antonsen (2009).

stating that 'we miss a great deal when we substitute power with culture' (Perrow 1999: 380).

Sagan (1993) extends and substantiates Perrow's emphasis on power from *Normal Accidents* by showing how a number of close calls involving nuclear weapons can be related to group interests and interest conflicts.

Perrow's message is clear. Organizational safety should be analysed through a power-oriented rather than a cultural perspective. While not dissecting Perrow's words in detail, I shall use his statement as a starting point in order to discuss the relationship between safety culture and the issue of power in organizations. This relationship is investigated by analysing the same case, the *Challenger* accident, through the lenses of both a power-oriented and a cultural perspective. The power-oriented perspective is informed by Lukes's (1974, 2005) three dimensional model of power, while the cultural perspective is informed by Vaughan's (1996, 1997) writings on the *Challenger* launch decision. My aim is to show that power-oriented theorists like Perrow are indeed correct in some of their criticisms of the existing research within the cultural approach. This does not mean that the cultural approach to safety cannot be fruitful but rather that the power-oriented criticism provides insights that can vastly strengthen research on and the understanding of safety cultures.

The Concept of Power: A Three-dimensional View

Power, like culture, is a concept notoriously hard to define. I shall make no attempt to explore the full depth of the conceptual labyrinths of power. A brief account of the dimensions of power will be sufficient for the argument presented here And to do so my account is based on Lukes's (1974, 2005) classification of power into three supplementary dimensions. Lukes's focus is on the general theoretical nature of power. While stating that power is a *capacity* (Lukes 2005), he does not say much about what may form the basis of this capacity, e.g. how power is created and re-created. This, in turn, is essential in order to analyse *which* actors or groups can be described as powerful. So I supplement Lukes's three dimensions of power with a description of the possible sources of power in organizations. The depiction of the sources of power is based on Pfeffer's (1981) account of power in organizations and Bolman and Deal's (2003) summary of the research on the political perspective on organizations.

The First Dimension of Power

Most theoretical accounts of the nature of power start off by citing Robert Dahl's classical definition, which states that 'A has power over B to the extent that he can get B to do something that B would not otherwise do' (Dahl 1957, cited in Lukes 1974:11). Dahl's definition is closely related to Weber's (1971) view on power as individuals' ability to carry out their will in a given situation.

Such definitions state that power is a sort of causal force that is sufficient to change other actors' actions but that nevertheless it is a characteristic of *relations* between social actors (Pfeffer 1981). It also follows that power is strategically exercised through intentional, rational calculation. The focus is on observable behaviour and processes of decision making. This constitute Lukes's (1974) first dimension of power, which may be based on one or more of the following sources:

1. *Position power.* Organizational structures distribute different levels of formal authority to the various positions in the hierarchy. In addition, the division of labour in an organization inevitably creates power differences since the various tasks and activities are not equally critical to organizational survival (Pfeffer 1981).
2. *Information and expertise.* This is the power of 'know-how' and 'know-what'. To control knowledge or information that is crucial to the organization is an important source of power. For instance, doctors are usually powerful in hospitals because they have a monopoly on the medical expertise that is required there. This is the sort of power that comes with being irreplaceable. In a similar vein, access to information that is important for the organization can also be a source of power. Some actors or groups hold positions that are central in the organizational information flow. These positions provide access to important information which can be used strategically.
3. *Control of rewards and resources.* Control over material, e.g. money and/or employment and immaterial, e.g. recognition, political support, resources are perhaps the most visible and common sources of power. Western companies often use monetary rewards to promote certain types of organizational goals like efficiency, service or safety. It is also quite common to reward behaviour by naming 'employees of the month' or with similar forms of immaterial rewards.
4. *Coercive power.* Coercive power is closely connected to actors' or groups' control over sanctions as it rests on 'the ability to constrain, block, interfere or punish' (Bolman and Deal 2003:195). Norwegian air traffic controllers' collective sick leave as a protest against cutbacks and organizational turbulence is a recent example of the exercise of this form of power. The air traffic controllers were able effectively to get their message across to Norwegian politicians, public, media and aviation authorities because of their high strategic importance in the business of aviation.
5. *Alliances and networks.* This is the power of 'know-who'. Network theorists, such as Granovetter (1973) have shown that alliances and networks can sometimes allow actors to tap into other actors' sources of power. The forming of coalitions consisting of different actors and groups looking to benefit from each other's power bases is the key interest of researchers analysing organizations as political systems.

6. *Personal power.* Charisma, energy, political skills and verbal facility are among the individual characteristics that constitute a source of power. In addition to these factors, Pfeffer includes a person's *reputation* for being powerful, smart, strong, etc. as an individually based source of power.

Lukes's first dimension encompasses the most visible form of power and indicates the more overt lines of conflict between actors and groups.

The Second Dimension of Power

Dahl's definition of power that forms the basis of Lukes's first dimension has been criticized for implying a rather mechanical model of agency and for overemphasizing the role of concrete decisions and decision-making (Clegg 1989). Critics like Bachrach and Baratz (1962), while not denying the importance of this form of visible power, argue that power also has another, more invisible, face. This is the face of non-decisions, referring to the ability to keep potential issues *out* of decision-making processes. This form of power is at work in situations where power struggle does not result in overt conflict but rather resembles a covert tug of war where groups or actors struggle to set the agenda for which potential issues are included in or excluded from the political processes that create decisions. This is Lukes's second dimension of power.

The sources of this form of power derive from differences between actors' access to and control of agendas. When organizational decisions are made the decision processes are always preceded by selection processes that single out *which alternatives* are worth considering as viable. This tends to give preferential treatment to the interests of those with a seat at the table, while the concerns of absentees are likely to be overlooked. This source of power has a close relationship to the alliances and networks of the first dimension. The building of alliances and forming of networks can provide access to the shaping of agendas for actors originally excluded from the process. In all cases, the second dimension of power gives rise to a far less obvious form of conflict than the first.

It is important to emphasize that the two dimensions of power introduced so far are not contradictory, but complementary. Lukes sees the development of a two-dimensional model of power as a significant advance in relation to a one-dimensional model consisting only of the form described by Dahl. But he still sees some fundamental shortcomings in a two-dimensional model, especially related to the way power is linked to the notion of conflict, whether overt or covert, and this forms the basis for the third dimension of power.

The Third Dimension of Power

In the third dimension of power, Lukes draws attention to the fact that not all instances of the use of power presuppose a conflict of interests. Moreover, not all exercise of power can be traced back to individually chosen acts. His point is that social systems tend to be biased in reflecting the values of a few groups at the expense of others. The creation and re-creation of this bias is neither consciously chosen nor the intended result of any individual's choices.

This form of power hinges on the construction of meaning in social life and enables the dominant to influence the dominated to adopt the goals, values and attitudes of the dominant. Lukes see this as a particularly effective form of power:

> Indeed, is it not the supreme exercise of power to get another or others to have the desires you want them to have – that is, to secure their compliance by controlling their thoughts and desires? (Lukes 2005: 27)

This may be labeled a cultural or symbolic form of power through its 'role in upholding one version of significance as true, fruitful, or beautiful against other possibilities' (Wolf 1994, in Alvesson 2002). Influencing the frameworks through which organizational members view reality can be a way to avoid argument and resistance by obscuring conflicts of interests. This represents a more systemic, Marxist, view of power. While the first two dimensions rest on a Weberian model of power, i.e. power as the ability to realize one's interests despite the resistance of others, the Marxist view on power emphasizes the way conflicts of interests can be obscured so that such resistance is eliminated or never created in the first place. In this outlook on power, the processes of socialization are central mechanisms that serve to favour the privileged. As I shall discuss in further detail below, this third dimension of power introduces a notion of culture into the study of power.

Lukes's point is that *all* these three dimensions are needed to cover the different aspects of power. They are not to be taken as competing theoretical frameworks but rather as necessary pieces in the puzzle of power. And this point is further reinforced by the fact that the sources of power overlap. For instance, positional power is very likely to imply also control over rewards; coercive power, access and control over agendas and perhaps also the opportunity to influence the meaning of symbols.

Lukes's model is summarized in Figure 4.1. Note how it is framed to show how the dimensions are to be conceived as layers of the concept of power.

The three dimensions of power I have delineated so far imply that power is relational in the sense that the concept refers to having power *over* someone. This view is inherent in both the Weberian and Marxist views, as they conceptualize power as a zero-sum game of constraining the will and actions of others. However, a quite different line of research emphasizes an *enabling* and more positive character of power; power as a resource to system integration (Law 1991). Power is not only the ability to restrict someone else's freedom, it can also constitute a form of freedom in itself. This is a more bottom-up view on power as it sees power as the creator of possibilities or initiating empowerment. When it comes to organizations, however, this view on power makes it difficult to talk about the *distribution* of power. Since the main focus of the 'power to' approach is on the

	Three-dimensional view		
	Two-dimensional view		
	One-dimensional view		
Key elements	First dimension	Second dimension	Third dimension
Object of analysis	- Behaviour - Concrete decisions - Issues	Interpretive understanding of intentional action Non-decisions Potential issues	Evaluative theorization of interests in action Political agenda Issues and potential issues
Indicators	Overt conflict	Covert conflict	Latent conflict

Figure 4.1   The three-dimensional model of power

Adapted from Lukes (1974).

role of power for social systems as a whole, the differences in power between actors or groups become less visible. In this work, therefore, while not denying the importance of power as 'power to', I shall place the main focus on the relational 'power over'.

Having delineated the three general dimensions of power and the more specific sources of power in organizations there is one remaining question. How does all this relate to organizational safety? I will try to answer this question by giving a brief account of some of the organizational processes that led to the accident with the space shuttle *Challenger* in 1986. Using *Challenger* as a case hardly qualifies for any awards for originality since it is probably the most well-documented, well-known and well-analysed accident within safety research. Nevertheless, the *Challenger* accident provides a good example because, uncommonly in accident research, it has been analysed *both* through power-oriented and cultural perspectives.

Power and organizational safety – The Challenger *launch*   The flight of the space shuttle *Challenger* ended in an explosive burn of hydrogen and oxygen propellants just 73 seconds after lift-off in January 1986. The triggering factor was a failure in the rubber O-rings that were supposed to seal the joints of the solid rocket booster (Rogers et al. 1986). It turned out that the O-rings lost their elasticity in the unusually cold weather that day. The problem with the O-rings was the topic of a teleconference the night before the launch, between NASA and the contractor responsible for the solid rocket booster, Morton Thiokol. The engineers of Morton Thiokol voiced their concerns that the combination of rubber O-rings and cold weather could severely threaten the safety of the mission and recommended delaying the launch. NASA officials at the Marshall Space Flight Center, under pressure to justify the huge costs involved with the space missions, responded quite harshly to this recommendation (Rogers et al. 1986). Faced with this reaction, senior managers at Morton Thiokol changed their minds and recommended the launch against the advice of the engineers. The Marshall Center never communicated the fact that the engineers had voiced concerns about the launch to higher levels of the NASA organization. Managers at the Marshall Center appeared to contain the problems, trying to deal with them internally, rather than communicating them to other parts of the organization (Rogers et al. 1986).

The processes leading up to this decision bear, in a number of ways, the imprints of power. First, NASA's push to go through with the launch must be seen in relation to its struggle for scarce resources allocated in the U.S government budget. NASA was behind schedule in delivering the planned number of flights and it seems highly likely that NASA officials feared that their credibility and future funding was at stake over the *Challenger* launch. This is an example of Lukes's first dimension of power: Congress has the ability to influence the actions of NASA through being able to control its financial resources.

Second, the relationship between NASA and Morton Thiokol is a relationship of power in that Morton Thiokol's existence is very much dependent on the NASA contract. NASA is thus in a position of power due to its control over the material resources, money and employment, on which Morton Thiokol is dependent. When NASA exercises this power it triggers the exercise of power within Morton Thiokol. When the Morton Thiokol managers, who are ultimately responsible for the survival of the organization, decide contrary to the concerns of their own engineers, it is a case of position power. By pulling rank, the managers exercised the power that comes with their positions in the formal hierarchy of the organization, i.e. the first dimension of power.

The report of the Rogers Commission places much emphasis on the form of power that corresponds to Lukes's first dimension: the open and visible power that manifests itself in the decision-making process leading up to the launch. However, it is also possible to identify traits of the power of non-decisions (Lukes's second dimension) in the Commission's findings. First, the way information was contained at the Marshall Center can be seen as an attempt to keep control over the decision making agenda regarding the launch. Then the engineer who voiced the highest concerns about launching in cold weather, Roger Boisjoly, felt he was sidelined in the concluding phase of the process:

> I was not even asked to participate in giving any input to the final decision charts
> ... I was never asked or polled, and it was clearly a management decision from
> that point. (Rogers et.al. 1986: 10)

Although Boisjoly states in the Commission's report that he feels he had his say in the process, the exclusion of his argument in the concluding phases of the decision making process illustrates that the opinions of the less powerful can sometimes be ignored when the real decisions are made.

The revelation of the overt power dynamics that surrounded the now infamous teleconference led both the media and the official investigators to conclude that the accident was caused by the managers' (mis-)use of power since they chose to disregard the protests from the engineers opposed to the launch. Perrow's comments on the *Challenger* accident further underline this, as he locates the accident's cause in 'the extraordinary display of power that overcame the objections of the engineers who opposed the launch' (Perrow 1999: 380). This comment pinpoints the strength of power and conflict oriented analyses of organizational safety. A clear picture of the framework of power is necessary to see how and why whistle-blowers are sometimes overruled when safety goals conflict with economic ones. This is probably why power oriented theorists like Perrow conclude that power is the key to understanding organizational realities and organizational disasters.

The Rogers Commission makes no attempt, however, to investigate the more invisible form of power, i.e. the way the third dimension of power contributed to the accident. This is not very surprising since the scope of official accident investigations is usually limited to investigating the immediate causes and decision making processes related to the accident sequence. The investigation nevertheless overlooks important factors contributing to the accident. In particular it fails to see the way the use of power was embedded in the cultural context of the NASA organization. The drive towards meeting the flight schedule significantly influenced which types of knowledge were considered valid, which safety margins were considered acceptable and, ultimately, which decisions were considered viable alternatives. In other words, the dominance of politicians and management, who insisted on meeting the flight schedule, influenced the values and meanings that constituted the NASA organizational culture.

The aim of this case study is to examine the relationship between cultural and power-oriented approaches to safety in organizational research. Therefore, I will elaborate further on this. I shall try to make a case for cultural approaches to organizational safety and show that culture and power are inextricably connected.

Culture, Power and Safety

The strength of power-based approaches lies in their emphasis on questions regarding whose interests organizations serve. Organizational culture is never politically neutral; it is likely to be biased in reflecting the values and world-views of dominant groups in the organization.

Perrow's scepticism towards the cultural approach, the starting point for this discussion, no doubt stems from the fact that much of the research on safety culture is based on an implicit model of organizational integration. Studying culture most often means searching for what is *shared* by organizational members, the basic assumptions, values, norms and knowledge that define membership of a social group or an organization (Schein 1992). The cultural perspective may thus be perceived to be directed more towards consensus than conflict even though this perception has been challenged also by proponents of an integrative definition of culture, see Schein 2004.

However, the argument that 'the issue is really power' has two shortcomings. The first has to do with the underlying assumptions about culture. The criticism seems to be based on an understanding of culture that is purely top-down, rationalistic and entirely integrative. This model implicitly assumes that people act quite passively in response to already existing cultural traits, a view which is very akin to the way organizational culture was described in the management literature of the early 1980s (e.g. Peters and Waterman 1983). This was a 'quick fix' view on organizational culture that has been heavily criticized for being far too optimistic about the chances of managers' being able to shape organizational culture at their discretion. If the management theories of organizational culture were correct in their depictions of culture, then Perrow's conclusion about the utility of the safety culture approach to safety would indeed be correct, as these theories sometimes give the impression that organizational culture has to do with employee brainwashing (Sejersted 1993).

But, as I attempted to show in Chapter 3, research and theory on organizational culture has advanced well beyond the simplistic managerialist view. There has been a shift in perspectives, from a functionalist towards a more constructivist view on culture which includes aspects of conflict and differentiation. Most researchers on organizational culture now see culture as both a product and a process. As well as structuring behaviour, culture is produced and reproduced through daily interaction (e.g. Meyerson 1991; Andersen 2001). It follows from this that organizational cultures are produced locally and that managers cannot expect to be able to design or modify them to suit themselves. In this way a constructivist perspective serves to moderate the power theorists' emphasis on the role of the powerful in defining reality. Less powerful actors cannot be seen as passive recipients of culture. This is illustrated by the fact that most organizations over a certain size have some local subcultures that may be in opposition to corporate management values and ideology. Therefore, while power relations no doubt *affect* the creation and re-creation of culture, culture is by no means *determined* by power relations.

The second shortcoming of the view that 'the issue is really power' is that it implies a kind of reductionism that boils the complexity of organizations and organizational safety down to a single factor. Placing all emphasis on the power perspective may lead one to overlook some important intersections between cultural and power-oriented approaches. After all, as is stated by Lukes's third dimension of power, the most efficient power is that which is hidden, i.e. not perceived as power by those being controlled. If the powerless do not even know they are being controlled, how is one to 'see' power? Power is often expressed in subtle ways, through the use of *symbols*. This means that the study of the least visible forms of power actually requires a cultural approach through the study of organizational symbolism.

The simplification and reductionism in Perrow's position have to do with the underlying theoretical assumptions of the power perspective. In order to show that a power perspective is not sufficient for the study of organizational safety, we need a more concrete example. Again, the *Challenger* accident can serve as an example.

Challenger Revisited

The immediate interpretation of the accident pointed to an abuse of power taking place the evening before the launch. This became subject to criticism a decade after the accident, especially through the writings of Diane Vaughan (1996, 1997). She challenged strongly the dichotomy of 'good' engineers versus 'bad' managers (Vaughan 1997:83) by looking beyond the infamous teleconference. When the decisions of the teleconference are placed as one in a long line of decisions, the story looks a bit different. Vaughan discovered that the problems with the O-rings, along with many other deviations from plan related to the rocket motors, were found in a number of missions prior to *Challenger*. These flaws were known to both Morton Thiokol and NASA, but were nevertheless allowed to recur for mission after mission. Vaughan's interpretation of this was that deviance had become normalized within NASA's organizational culture. The pressure of cost effectiveness influenced the organizational culture by pushing the boundaries of what was regarded as acceptable risk. This process has later been labelled 'practical drift' by Snook (2000).

Another interesting point in Vaughan's 1997 analysis regards the occupational identity of the engineering profession. Both NASA and Morton Thiokol were characterized by a technical culture where positivist rationality was a central ideal. Risk assessments and decisions were to be based on rigorous quantitative analysis and repeated testing. Hunches or intuitions had no place in the processes of making launch decisions. This cultural ideal contributed to the NASA officials' negative response to the engineers opposing the launch in the teleconference because to a large extent their opposition was based on intuition rather than test data.

Vaughan's 1997 interpretations differ markedly from the power oriented interpretations expressed by the Rogers Commission and Perrow (1999). Instead of focussing on the concrete conflicts of the teleconference, she makes an argument focussing on the social construction of reality. The process of social construction can explain how, over time, the practices of both NASA and Morton Thiokol led to the normalization of deviance. The long-term effects on an organization of such socially constructed practices is illustrated by the fact that many of the causes behind the *Challenger* accident were also present in the *Columbia* accident seventeen years later (CAIB 2003).

My point here is that Vaughan's analysis captures some aspects that the power-based analysis overlooks. When taking into account the cultural settings of the decision-making process, she is able to show how the decision to launch could seem appropriate for the actors involved at the time. The study of organizational power and politics focuses on decisions with an aim to describe who gets what and how they get it (Pfeffer 1981). From a cultural perspective, the question is not who made a decision and how, but *why* they made that particular one.

Having said this, however, what stands out is how little Vaughan's analysis emphasizes the relationship between NASA and Morton Thiokol. Her analytical focus seems to be on the historical and cultural context of the decision-making processes, but she really does not deal with the asymmetrical relationship between the two organizations involved.

She does to some extent acknowledge the role played by political and competitive factors as 'structural conditions' (1996: 198) for the culture of production. In this way she includes a notion of Lukes's first dimension of power based in the control over resources. By including this institutional dimension in the analysis, Vaughan thereby illustrates that intra-organizational dimensions of safety are influenced by external factors. What she does not do, however, is discuss how these conditions affected the construction of what was regarded as acceptable risk. This is a question of Lukes's third dimension of power.

Vaughan's account of *Challenger* seems to see the social construction of reality as a quite democratic process albeit within the context of structural conditions. According to Lukes's third dimension of power, however, this would hardly be the case. The normalization of deviance, for instance, does not come about in a power-free context. As Vaughan states herself, NASA barely managed to get the shuttle programme endorsed after the Apollo era and only did so by 'top managers ... selling the shuttle to Congress as a project that would, to a great extent, pay its own way' (Vaughan 1997: 89). In other words, the top administrators of NASA had positioned the organization as becoming more like a commercial organization and committed themselves to achieving this goal. The culture of production that characterized the NASA organization was no doubt affected by this. Thus, the normalization of deviance did not come about as a result of everyday practice alone, but was also result of a system bias in the NASA organization that was introduced by powerful forces. The organizational culture was shaped by the conditions laid down at the top of the organization. To use the words of the anthropologist Bob Scholte, while culture consists of webs of significance that actors spin themselves, 'very few do the actual spinning, while the majority is simply caught' (Scholte 1984, cited in Keesing 1987: 162).

Also, the pre-eminence of the technical culture in NASA can be seen as a form of cultural domination. It represents a shaping of the ground rules of the organization, of what is regarded as a true and valid basis for communication and decision making (Alvesson 2002). On the one hand, the belief in scientific objectivity and rationality are forms of knowledge that exercise power on both managers and engineers because they are a prevailing cultural ideal in western societies (Foucault 1980). On the other hand, the way the advice of the engineers at Morton Thiokol was disregarded by managers illustrates that management may have a privileged position in shaping the frames of reference through which organizational activity is seen and evaluated. In my view, both these forms of cultural domination are related to Lukes's third dimension of power.

The conclusion to be reached after revisiting the *Challenger* accident is that Vaughan's analysis captures some aspects that the power-oriented perspective overlooks. But the power-oriented criticism of the cultural approach is, however, justified on several grounds. When faced with such criticism, safety culture researchers should not be dogmatic in defending their own approach but rather profit from the critical comments to improve their own research. In the following paragraphs I will try to sketch out the lessons to be learnt from the power-oriented critique of safety culture research.

What are the Lessons for Safety Culture Research?

The most obvious but still the most important lesson for safety culture research is the fact that organizations are arenas for conflicting interests. Safety issues, like any other organizational issues, are subject to discussion and disagreement. For one thing there is often disagreement about what is dangerous and what is safe. The infamous teleconference the night before the *Challenger* accident is one example of this. A more recent example is a controversial issue in the Norwegian oil industry where a group of North Sea workers experienced deferred injuries allegedly due to exposure to oil fumes. While workers and medical personnel had complained about nausea and headache for decades, the occupational hygienists reportedly responded by asking them to 'stop hassling about the oil fumes; the fumes are not dangerous' (Dagbladet 2005). Although this is a journalistic report and not research, the workers' experience of being disregarded serves as an example of how risk is socially constructed and how its definition is influenced by power.

Acknowledging the role of power and politics in organizational life does not mean, however, that we can discard the integration perspective. The aim of the study of culture should still be to examine the shared frames of reference through which action strategies are constructed (Swidler 1986). The argument is rather that in addition to asking what is shared, it is important to ask what is *not* shared. The inclusion of this dimension enables the cultural perspective to analyse conflict both within a culture and between cultural units.

The fact that power differences, conflict and differentiation are rarely discussed in safety culture research is somewhat anomalous since such perspectives have been influential within organization studies for half a century. Indeed, neglecting power dynamics in safety culture research is the more surprising since the treatment of whistle-blowers, which is at the core of the concept of safety culture (cf. Westrum 1993), is almost by definition associated with power relations.

It is worth noting that the treatment of whistle-blowers by organizations lies at an intersection between cultural and power-oriented approaches to organizational safety. For instance, both the power-oriented and the cultural explanation of the

Challenger case focused on how the engineers who opposed the launch were treated by their superiors. While the power-oriented explanation aimed at singling out the power relations present in the process, the cultural explanation aimed at describing the cultural settings that *legitimized* the exercise of power, notably the technical culture with its preference for hard quantitative data over hunches and gut-feeling. This illustrates that although they explore two different levels in the causes of accidents, issues of power and of culture are inextricably intertwined. The one analyses the decision-making processes that produced the accident while the other analyses the underlying cultural settings that provided grounds for the decision making processes.

What this adds up to is a broadening of the concept of safety culture that has several advantages over a purely integrative approach. In particular, a more conflict oriented approach is better at analysing instances of worker resistance through the formation of informal workers' collectives (Crozier 1964; Lysgaard 2001). The potential friction between management and shop floor subcultures is highly relevant for the study of safety since it is likely to affect the endorsement and thus the proficiency of safety management systems, as well as reveal the conditions that create whistle- blowers at shop floor level.

A power-oriented view on safety culture could also be instrumental in breaking the ground for a critical examination of the underlying theoretical and philosophical assumptions of the concept of safety culture. As I will discuss in more detail in Chapter 7, such a critical reappraisal would seriously question some of the existing approaches to safety culture improvement, especially those that aim to change culture through attitude modification and the principles of Behaviour Based Safety (BBS). Such improvement programmes have become increasingly common, especially in the petroleum industry. To a great extent these approaches equate safety culture with following rules and procedures. The measures which are implemented to achieve this include discipline through reward or punishment, attitude campaigns and training and educational programmes.

The strategy can be summed up as attempts at getting the 'right' results through establishing the 'right' culture. From a power-oriented perspective this can be seen as an example of power trying to control employees' construction of meaning. As such, this corresponds to Lukes's third dimension of power. The desire to measure, monitor and ultimately change the workers' culture can also be seen as an example of what Foucault (1977) calls the spreading of 'prison-like' surveillance, control and discipline into more and more areas of human activity. A top-down driven model of safety culture expands the scope of safety management from initially aiming at controlling the actions of employees to wanting control over their hearts and minds.

Clearly ethical considerations apply to this approach. There is an inherent paradox in the 'quick fix' literature on organizational culture, which requires managers to:

brainwash their employees and, at the same time, treat them as individuals. There is no discussion of the possible antagonism between these two strategies (Sejersted 1993:2).[2]

This quotation illustrates what can be seen as a basic manipulative attitude inherent, but unspoken, in some management theories. The pursuit of consensus, which characterizes many management theories, can be ethically problematic because building consensus can easily turn into manipulation.

As much safety culture research falls within the integrative paradigm, and is based on such a consensus-oriented view of organizations, such ethical reflections are highly relevant for safety culture research. It is a pity that they are virtually nonexistent in the literature on safety culture and emphasizes that a critical reappraisal of the theoretical and ethical grounds of the field may be long overdue.

Safety Culture and Power – Some Concluding Remarks

The aim of this chapter was to discuss the relationship between power-oriented and cultural approaches to organizational safety and was achieved by using the accident with space shuttle *Challenger* as a case because this accident has been analysed through both approaches. I have argued that issues of culture and power are so intertwined that the concept of power should be included in the study of safety culture. Incorporating Lukes's three dimensional model of power would be a useful way to improve the cultural approach to safety.

Relating safety culture to power implies moving beyond some of the traditional conceptions of safety culture where culture is associated with consensus and harmony. This is quite necessary both to be able to give more valid accounts of organizational dynamics regarding safety but also in order to avoid the mutation of the concept of safety culture into authoritarian safety doctrines which would fail to preserve the scientific value of the concept of safety culture.

---

2   My translation.

Chapter 5
Assessing Safety Culture

The interest in safety culture is based, at least to some extent, on a hypothesis that there is a connection between the cultural traits and the level of safety in an organization. Descriptions of an organization's safety culture are considered to provide grounds for proactive safety management by yielding 'predictive measures ... which may reduce the need to wait for the system to fail in order to identify weaknesses and to take remedial actions' (Flin et al. 2000: 178). The hope is that such 'predictive measures' will constitute leading, as opposed to lagging, indicators for the management of safety. All of this depends on the ability reliably to assess and monitor the relevant factors that cause accidents.

In safety culture research, questionnaire surveys appear to be the predominant strategy for such assessment and monitoring (Hopkins 2006). For instance, all but one of the seven empirical articles in *Safety Science*'s special issue on safety culture[1] employed survey methods. Some authors have argued that questionnaire-based methods should be reserved for studies of safety *climate*, which is perceived as a more superficial phenomenon than culture (e.g. Cox and Flin 1998; Hopkins 2006). However, survey methods are still regarded as a satisfactory way to study attitudes, values and perceptions about organizational practices (Hopkins 2006) and as a valid means of providing a snap shot of the current state of safety (Tharaldsen et al. 2008: 3). There are also several recent empirical studies studying safety culture by means of survey methods (e.g. Ek et al. 2007; Fernández-Muñiz et al. 2007). In spite of there being some disagreement as to whether the aspects studied are to be labeled culture or climate, I choose to use the term culture. The reasons for this are firstly, that the trend is for the concept of culture to gain prominence at the expense of the concept of climate and secondly, that there are no clear conceptual differences or other distinctions between studies of climate and studies of culture (Reichers and Schneider 1990; Cox and Flin 1998).

The Limits of Survey Methods – A Case Study

If survey methods are considered to be useful towards assessing safety, the results of such assessments should provide some basis for making judgements about how safe or unsafe an organization is, as well as some sort of 'prediction' as to whether the organization is likely to have accidents in the future. The predictive value of culture surveys has not been the subject of much discussion or empirical

---

1   *Safety Science* 34 (1–3), 2000.

investigation in the research literature. Although some authors have addressed the topic (e.g. Cooper and Phillips 2004; Hofmann and Mark 2006; Mearns et al. 2003; Zohar 2000), knowledge about the relationship between culture surveys and safety is rather limited. The existing research also largely focuses on the relationship between safety culture measures and *occupational* accidents. The predictive value of culture surveys in relation to major *organizational* accidents has, to my knowledge, never been empirically analysed.

The case study targets this research gap by presenting an empirical analysis which presents, and in some areas also questions, the usefulness of survey methods as a tool for safety culture assessment. It should be stressed that the survey was not designed to predict the likelihood of a major accident, so the case study is not intended to take an 'Aunt Sally' approach where the questionnaire study is criticized for its failure to meet its intentions. Nevertheless, the very utility of the concepts of safety climate and safety culture rests on an assumption that they have something to say about the safety level in an organization. After all, if assessments of safety culture and climate were not regarded as having anything to say about the conditions for safety in organizations, why should we bother conducting them in the first place?

The data material consists of a safety culture survey conducted on the Norwegian oil and gas installation Snorre Alpha in 2003 and the reports of two investigations into a very serious incident on Snorre Alpha in 2004. The existence of two independent sources of scientific data,[2] one collected before a serious incident and the other after it, makes Snorre Alpha a unique case for studying the relationship between these two types of data. As the investigations into the incident revealed several problems that can be related to culture/climate, Snorre Alpha constitutes what is commonly labelled a 'most likely case' (Eckstein 1975) to study the link between proactive and reactive assessments of the conditions for safety. If there is indeed a relationship between survey assessments of safety culture and the 'real' conditions for safety in an organization, then this should be evident in the comparison between the proactive and reactive assessment of the safety situation on Snorre Alpha.

As the remainder of this chapter will show, there is a considerable gap between the situation depicted by the culture survey and the situation described in the accident investigations.

Gas Blowout at Snorre Alpha

Snorre is an oil field in the Tampen area in the northern part of the North Sea. The sea depth in the area is 300–350 metres. The oil on Snorre is produced with pressure maintenance by water injection, gas injection and water alternating gas.

---

2   One of the investigations was led by Professor Per Morten Schiefloe at the Norwegian University of Science and Technology. Since the investigation was set up much like a research project, the data investigation is seen as a source of scientific data.

Snorre Alpha is a floating steel facility for drilling and processing as well as providing accommodation for 220 people.

On Sunday 28 November 2004, an uncontrolled situation occurred during work in Well P-31A on the Snorre Alpha facility. The operation to be performed is called slot recovery. It consists of pulling pipes out of the well in preparation for the drilling of a sidetrack. During the course of the day the situation developed into an uncontrolled gas blowout on the seabed, resulting in gas under and on the facility. Of the 216 people on board the platform at the time of the blowout, 181 were evacuated by helicopter to installations nearby. The 35 people left on the platform tried to gain control of the situation by pumping drilling mud into the well, a process known in offshore terminology as 'bullheading'. This work was complicated by the fact that there was gas under the installation, which made it impossible for any vessels to supply additional drilling mud since the proximity of a running combustion engine could have ignited the gas. During the night, the installation's supplies of mud ran out and the only choice was to mix mud out of available drilling fluids, manually. This approach was successful, and the well was stabilized on the 29 November.

The actual consequences of the incident are mainly related to disturbances in production and the economic losses connected with this. Production on Snorre Alpha and the neighbouring platform Vigdis is about 200 000 barrels of oil per day.[3] Production on both platforms was stopped by the incident and did not return to normal for several months afterwards (Brattbakk et al. 2005). No one was killed or injured as a result of the incident. Under slightly different circumstances, however, it could have led to the loss of many lives. The gas could have ignited, causing a scenario similar to the Piper Alpha accident (cf. Cullen 1990). The quantities of gas flowing from beneath the platform could also have resulted in stability problems which, in turn, could have made it capsize. Both these scenarios imply the loss of many lives, environmental damage and additional loss of material assets. Owing to the serious potential of the incident, the Norwegian Petroleum Safety Authority (PSA) has characterized the incident as one of the most serious ever to occur on the Norwegian shelf (Brattbakk et al. 2005).

Methods and data Finding appropriate ways to 'measure' different aspects of culture has been a recurrent problem for both practitioners and researchers interested in safety culture. Several researchers and institutions have developed questionnaires containing items meant to be indicators of cultural traits regarded as important for safety. Examples of existing survey tools include the 'Health and Safety Survey Climate Tool' (HSE 2001), the 'Safety Culture Assessment Toolkit'[4] (Cox and Cheyne 2000) and 'Risk Level on the Norwegian Continental Shelf'

---

3   The oil price at the time of the accident was around 50 USD per barrel. The incident thus caused economic losses in the region of ten million dollars per day.

4   The Safety Culture Assessment Toolkit also includes qualitative techniques and provides guidance in the use of these.

(Tharaldsen et al. 2008). The hope is that the use of such assessment tools will provide the basis for a proactive analysis of risk.

The need for such proactive assessments has been emphasized by high reliability theorists (e.g. Rochlin et al. 1987) and more recently by Hollnagel et al. (2006) in their highly influential book *Resilience Engineering*. According to Hollnagel and colleagues, the first step towards building resilience is 'to analyse, measure and monitor the resilience of organizations in their operative environments' (Woods and Hollnagel 2006: 6). This relates to organizations' ability to recognize when situations fall outside the limits of acceptable risk and thus to the organizational foresight that is an important part of resilience engineering. How one is to analyse and monitor the social part of socio-technical systems is, however, not explicitly discussed. Since the practical value of safety monitoring rests on its ability to describe current and 'predict' future performance, this is a very important question for safety research.

In recent years, there have been some signs that qualitative research may be gaining greater prominence in safety culture research. Authors such as Richter and Koch (2004) and Haukelid (2008) have introduced ethnographically inspired perspectives to the study of safety culture. Some of the quantitatively oriented researchers are also starting to question the usefulness of survey methods. For instance, Guldenmund (2007) suspects that safety culture questionnaires 'invite respondents to espouse rationalisations, aspirations, cognitions or attitudes at best', and that we 'simply do not know how to interpret the scales and factors resulting from this research' (Guldenmund 2007: 727). Such suspicions form the backdrop of the analysis presented in this case study.

The question to be addressed is the degree to which safety culture surveys predict whether an organization is likely to experience major accidents. This question is answered by comparing the descriptions of an organization's practices before and after a major incident. As already indicated, the organization in question is the oil installation Snorre Alpha.

To a large extent the analysis is based on the comparison of existing data which are described in three reports. The results from the culture survey conducted prior to the incident are described in two reports by Kongsvik (2003) and Antonsen and Kongsvik (2004). The incident was analysed both by the Petroleum Safety Authority (Brattbakk et al. 2005) and a joint group of Statoil representatives and researchers at the Norwegian University of Science and Technology (Schiefloe et al. 2005). In addition to the results described in these reports, some additional analyses are made in order to control the validity of the culture survey. Before proceeding to the analysis, a description of the methodical basis of these sources of data is in order.

The survey

The employees on Snorre Alpha completed a self-administered questionnaire consisting of 20 items before they participated in a company-wide programme for improving operational safety (the Safe Behaviour Programme) in 2003. The survey was designed and analysed by researchers from the same institute that participated in the causal investigation after the incident.[5] The questionnaire was developed to measure key aspects of safety culture and was assessed and revised by a group of Statoil representatives with good knowledge of the target group. The questionnaire is further described in Kongsvik (2003) and Antonsen and Kongsvik (2004). The respondents received the questionnaire and information about the survey by mail two to three weeks before the safe behaviour programme. They were asked to complete the questionnaire and enclose it in an envelope to be handed over as they registered for the Safe Behaviour Programme.

The questionnaire included quite common safety culture indicators related to dimensions such as managers' prioritization of safety, safety communication, individual risk assessment, supportive environment and safety rules and procedures, see Table 5.1 for the questionnaire items. Although the selection of questionnaire items will always be subject to disagreement among researchers, those included should be expected to yield information on organizational aspects that are highly important for safety.

The items were rated on a six-point Likert scale, ranging from 'totally disagree' to 'totally agree'. Four hundred and fifteen employees completed the questionnaire, an 82.5 per cent response rate. Of these, 197 were employed by Statoil, 215 by various contractors and three people did not disclose their employer. All three shifts working on the platform were included in the survey.

For the purpose of this study, there is little need for any advanced analyses of the data from the culture survey. An account of it will be largely sufficient in order to compare the two descriptions of the Snorre Alpha safety culture. However, some additional analysis is necessary. As already indicated, the comparison between the before and after situations reveals considerable divergence. In order to investigate the possible explanations for this, two additional analyses of the culture survey data are included to explore whether the results of the culture survey could have been influenced by either the characteristics of the questionnaire used or the characteristics of the population involved, i.e. whether the employees at Snorre Alpha differ significantly from employees at other North Sea installations.

The first reason for the divergence between the analyses made before and after the incident could have been due to properties of the questionnaire used in the analysis of Snorre Alpha. Incidentally, the oil company that financed the study wanted to try an alternative questionnaire design for some of their other installations, so the questionnaire used in the analysis of Snorre Alpha was modified. In the initial questionnaire all items were formulated as statements but were reformulated into questions, the hypothesis being that items formulated as questions would be more neutral than value-laden statements (Dallner et al. 2000). An illustrative example would be: The statement 'I always consider the risks involved before I carry out my work', was rephrased to 'Do you consider the risks involved before you carry

---

5   The author of this book is employed by the same institute but was not involved with the survey at Snorre Alpha or the causal investigation.

out your work?'. This may seem like splitting hairs, but some researchers argue that statements like the original one have an inherent positive charge. To disagree with such a statement may conflict with the respondent's image of himself/herself as a good employee. Of course, this conflict may also be present when the item is phrased as a question but the assumption is that questions avoid at least some of it (Dallner et al. 2000). The alternative survey was carried out on three oil installations when personnel attended the company's Safe Behaviour Programme. In this survey 144 people participated with a response rate of 62 per cent.

The second reason for the divergence could be attributed to some special characteristics of Snorre Alpha personnel. Therefore, the results from the survey at Snorre Alpha will be compared with the results of the employees at the three other installations. The difference between Snorre Alpha employees and those working on other installations was tested using a t-test for equality of means. This survey was answered by 1131 respondents (response rate 83.7 percent).

The incident investigation and causal analysis
The incident on Snorre Alpha was, of course, investigated by the Norwegian Petroleum Safety Authority. The investigation group conducted interviews with relevant personnel, studied documents that related to the incident as well as conducted on-site inspections of the platform (Brattbakk et al. 2005). Based on MTO analysis,[6] the aim of the investigation was to map the course of events, identify possible breaches of regulations and recommend actions to prevent future incidents (ibid.). The investigation by the PSA aimed to uncover the triggering and root causes of the blowout. It ended with, among other things, an injunction ordering Statoil to investigate further into the root causes of the incident. This formed the background for a 'causal analysis'[7] undertaken by the joint group of researchers and Statoil representatives.

The causal analysis was a very thorough study. In addition to conducting 152 interviews with onshore and offshore employees, a wide range of documents, HSE statistics, reported incidents, internal studies of working environment was studied and a questionnaire study undertaken (Schiefloe et al. 2005). The interviews took informants from all parts and levels of the onshore and offshore organizations and included not only Statoil employees but also from the drilling contractor and other contracting companies. The interviews included people who were both directly involved in the incident and those who worked on other shifts and were conducted between the end of June and the end of September 2005. Three visits were made to Snorre Alpha and 61 of the interviews were conducted offshore (Schiefloe et al.

---

6   Man, Technology and Organization.

7   The term 'causal analysis' is a direct translation from the Norwegian '*årsaksanalyse*'. The term 'causal analysis' refers to a broad investigation into the various organizational, cultural, technological, relational and interactional factors seen as contributing to the incident. The notion of causality should therefore not be taken in a strict sense, as implying any form of law-like relationships.

2005). The study was based on an extended MTO analysis, where the analysis of the human contribution is broadened to include safety culture and other aspects of informal organization.

The two reports described here form the basis of the post-incident description of the safety culture on Snorre Alpha. The incident investigations have a somewhat wider scope than the culture survey, as they include personnel in the onshore organization in addition to Snorre Alpha employees. This difference in scope is not a problem for the comparison made here, as the findings in the investigations are attributed equally to onshore and offshore personnel (Brattbakk et al. 2005)

Before the incident – culture survey results   The purpose of this study is to compare two different sources of data regarding the safety situation on the North Sea oil platform Snorre Alpha. The comparison between the results of the safety culture survey and the findings of the investigations after the gas blowout will shed light on the predictive value of the safety culture survey.

The safety culture survey results are described in a report by Kongsvik (2003). The mean values and frequency distributions of the analysis are presented in Table 5.1. The main trend in the data is predominantly positive.

Of all 20 items, only two have a mean score below 4.5. Considering that the Likert scale used ranges from one to six, these are undoubtedly positive scores. The item most negatively evaluated is '*My supervisor often calls on me at my working place to discuss safety*'. Almost 45 per cent of the respondents are found in the three lower categories. Also, the item relating to the company's expressed value that 'we always have the time to work safely' scores somewhat lower than the main tendency of the data material.

On the other side of the scale, seven of the items had a mean value of 5.0 or higher. Among these are statements concerning compliance with procedures, prioritization of safety and organizational learning, all considered key aspects of safety culture. Another positive finding of the survey analysis was that the vast majority, around 95 per cent, of respondents state they consider the risks involved both prior to and during work operations. As we shall see, this was one of the areas where the incident investigation and causal analysis came to opposite conclusions.

The overall conclusion of the survey results is, nevertheless, relatively clear: 'altogether, the employees' evaluation of aspects of the safety culture is highly positive' (Kongsvik 2003: 24; my translation). The presumed proactive assessment thus leads to the conclusion that – almost – all is well with regard to the Snorre Alpha safety culture.

At this point it is necessary to make a short detour from the main argument. After all, how can we be sure that the positive results of the Snorre Alpha survey are not due to some measurement error? We therefore need to make some additional analyses regarding the validity of the culture survey.

Table 5.1   Mean values and frequency distribution (percentages) of items of the safety culture survey

Questionnaire Item	Mean	1 (%)	2 (%)	3 (%)	4 (%)	5 (%)	6 (%)
My supervisor sets a good example when it comes to safety at my workplace	4.7	0	1	9	26	51	13
When somebody at my workplace brings forward safety-related information, the unit manager will make sure the problems are solved	4.8	0	1	7	24	48	20
Management will follow up on actions from HSE-inspections and meetings	4.7	0	1	8	33	40	18
My supervisor appreciates my coming forward with safety-related issues	5.0	0	1	4	16	49	30
My supervisor often calls on me at my working place to discuss safety	3.6	5	14	25	34	17	5
In our organization it is common to intervene if someone works in a hazardous way	5.0	0	2	5	20	44	29
We show care for each other in our daily work	5.1	0	1	2	18	48	30
I always consider the risks involved before I carry out my work	4.9	0	0	5	25	42	28
I continuously consider the risks involved while I carry out my work	4.9	0	0	4	22	50	24
Improving safety has a high priority at my workplace	5.0	0	1	4	20	47	28
The principle that 'we always have the time to work safely' is lived up to at my workplace	4.4	1	3	12	34	34	16
At my workplace, work operations are always stopped if there are any doubts as to whether safety is ensured	4.8	0	2	4	26	42	22
Our managers will take action if safety measures are not implemented within given deadlines	4.5	0	4	10	31	40	15

Table 5.1 *Concluded*

Questionnaire Item	Mean	1 (%)	2 (%)	3 (%)	4 (%)	5 (%)	6 (%)
At my workplace, operations that involve risk are carried out in compliance to rules and regulations	5.0	0	0	5	18	50	27
Injuries and near misses are always reported in accordance with regulations	4.9	0	1	7	25	37	30
At my workplace, deliberate breaches of rules and regulations will always be sanctioned	4.9	0	0	7	25	42	26
When undesirable events happen at my workplace, measures will be taken to prevent similar incidents from happening in the future	5.0	0	1	5	20	46	28
If I make a mistake, I can report it to management without fear of negative reactions	4.8	0	2	9	19	43	27
I never refrain from reporting undesirable events out of fear of negative sanctions from my co-workers	4.6	2	4	7	26	41	21
All in all, I think we have a good safety culture at my workplace	5.0	0	1	2	15	58	24

1= strongly disagree, 2= disagree, 3= somewhat disagree, 4= somewhat agree, 5= agree, 6= strongly agree.

Exploring the validity of the culture survey   The results from the culture survey could conceivably be influenced by measurement errors related to characteristics of the questionnaire. This possibility must be explored before coming to any conclusions about its predictive value.

If the positive results from the survey on Snorre Alpha are to be attributed to properties of the questionnaire it would have to be related to one of two issues. Either the questionnaire asks the wrong questions or the questionnaire items are formulated in the wrong way. And there may also be the possibility that the results of the survey are due to some characteristic of the work force on Snorre Alpha. All these three elements might have influenced the results of the survey, so I shall now examine each of these possible sources of 'noise'.

Asking the right questions?

This issue is a theoretical question concerning which dimensions are thought to be of relevance to safety culture. For instance, some might perhaps object that the questionnaire lacks items regarding training, job satisfaction and organizational interfaces, or that it is somewhat narrowly focused on individual behaviour. However, when compared with the theoretical dimensions of safety culture, as described by, for instance, Reason (1997), Guldenmund (2000) or Hale (2000) it seems that the questionnaire covers most of the dimensions mentioned. Also, when compared with existing questionnaires such as the Health and Safety Climate Survey Tool (HSE 2001) and the Safety Climate Assessment Toolkit, abbreviated SCAT, cf. Cox and Cheyne (2000) it appears that the questionnaire used in the study of Snorre Alpha includes similar items. Compared with the SCAT survey, for instance, both questionnaires include very comparable items about management prioritization of safety, safety-related communication, safety rules and procedures as well as the degree of social support among colleagues. There are some differences between the two questionnaires; the Snorre Alpha one places more emphasis on individual risk assessment while the SCAT one includes several items regarding personal appreciation of risk and the level of worker involvement regarding safety issues (SCAT, undated). Despite these differences, the key dimensions of the questionnaires are alike – although the words used may be somewhat different, the theoretical content is pretty much the same.

Another basis of comparison is found in Håvold's doctoral thesis on the 'measurement' of safety culture (Håvold 2007), where he reviewed the state of the art on culture surveys. He showed that that the five most common factors found in quantitative studies of safety culture surveys were safety rules, management commitment to safety, safety behaviour, communication, and work situation/ work pressure. This list of factors largely corresponds to the factors found in a similar review conducted by Flin et al. (2000). The Snorre Alpha questionnaire included items related to all of the recurring themes identified by these systematic reviews.

In other words, while one could always debate the selection and wording of items, there is no reason to believe that the questionnaire used in the study of Snorre Alpha was off target in regard to the themes included in the survey.

Asking in the right way?
The question of whether the phrasing of the items had influenced the results of the survey was empirically tested by Antonsen and Kongsvik (2004), who reformulated the items in the questionnaire from statements to questions. The hypothesis was that phrasing the items like questions would avoid the bias inherent in positively laden statements.

The comparison of the results from Snorre Alpha with the results of the study based on the alternative questionnaire design reveals that there were few differences between the two. In fact, the study with the alternative design yielded somewhat higher average item ratings – the standard deviations were somewhat greater for the items phrased as questions. Admittedly, the comparison of the two questionnaire designs may have a weakness in that they were administered to two different populations. However, using two different samples avoids the risks of maturation effects that could have influenced the results had the same respondents been used. Also the two populations involved are quite similar in terms of work content and context, educational background and regulatory regime. So the comparison is considered valid and therefore we can conclude that the comparison of the two questionnaire studies does not support the hypothesis that the positive description of Snorre Alpha is attributable to characteristics of the questionnaire design.

Snorre Alpha compared to other installations
We need to consider whether the positive results from the culture survey could also be related to some special characteristics of the work force on Snorre Alpha. For instance, it could be that the results were even more positive on other installations, so the questionnaire in fact did point to problems on Snorre Alpha when compared to other installations. In order to investigate this possibility, the results of the Snorre Alpha survey were compared with the results obtained from several other North Sea oil installations operated by the same oil company as Snorre Alpha. The survey on the other installations used the same questionnaire as Snorre Alpha.

The results show that responses on the included indicators from Snorre Alpha do not differ radically from those from the other installations. The items rated the least positively by the Snorre Alpha employees are also the least positively rated on the other installations. However, there are a few statistically significant differences between Snorre Alpha and the other installations and the items where these differences arose are presented in Table 5.2.

This comparison shows that Snorre Alpha scores are somewhat lower on three items related to management, as well as the overall evaluation of safety culture. These differences are statistically significant at the 0.05 level of significance using a t-test for equality of means. However, the sample sizes are rather large, so one

Table 5.2   Mean values and frequency distribution (percentages) of items with statistically significant differences between Snorre Alpha and three other installations

Questionnaire Item	Installation	Mean	1 (%)	2 (%)	3 (%)	4 (%)	5 (%)	6 (%)
My supervisor sets a good example when it comes to safety at my workplace	Snorre Alpha	4.7	0	1	9	26	51	13
	Other	4.8	0	2	6	24	50	19
Management will follow up on actions from HSE-inspections and meetings	Snorre Alpha	4.7	0	1	8	33	40	18
	Other	4.9	0	1	5	23	49	22
My supervisor often calls on me at my working place to discuss safety	Snorre Alpha	3.6	5	14	25	34	17	5
	Other	3.8	6	10	21	35	23	6
All in all, I think we have a good safety culture at my workplace	Snorre Alpha	5.0	0	1	2	15	58	24
	Other	5.1	0	0	1	12	57	30

1= strongly disagree, 2= disagree, 3= somewhat disagree, 4= somewhat agree, 5= agree, 6= strongly agree.

would expect some statistically significant differences in a sample of 20 items. Despite being statistically significant, therefore, these differences cannot be seen as representing any major differences between the two groups.

The main conclusion from this detour is that there is little reason to believe that the highly positive results from the Snorre Alpha culture survey can be attributed to some population characteristic, or characteristics of the measurement instrument. Thus, the results of the survey can be said to be valid.

The overall image of safety culture that can be perceived from the survey results is of a culture that is one of compliance and learning, sensitive of the risks involved and highly oriented towards safety, almost resembling what has been labelled a 'generative' culture by Westrum (1993) and Lawrie et al. (2006).

In the following sections this description of the Snorre Alpha safety culture is compared with the findings of the two post-incident analyses. As will be shown, the post-incident description is quite different.

After the incident – investigation and causal analysis   The Petroleum Safety Authority's official investigation of the incident (Brattbakk et al. 2005) identified a total of 28 departures from the regulations that set out the demands of technical, operational and organizational safety barriers. In particular, there were severe deficiencies in the planning and accomplishment of the well operation. According to the report, these deficiencies related to the following issues:

1. A lack of compliance with procedures and governance documents. According to the PSA, 'the lack of compliance was evident in all phases of the operation, but in particular in the planning of the operation' (Brattbakk et al. 2005: 23; my translation).
2. A lack of understanding of risk assessments and deficiencies in carrying them out. According to the PSA, 'the investigation shows that risk assessments are given low priority[8], there was a lack of holistic understanding of the risks involved, and in one instance risk contributions had been removed from the detail programme' (Brattbakk et al. 2005: 23; my translation).
3. Insufficient managerial involvement. In particular, the PSA emphasizes the 'insufficient managing of resources' in planning the operation, the failure to 'involve [external] competence units necessary to reveal the deficiencies related to risk assessments and training in the use of procedures' (Brattbakk et al. 2005: 24; my translation).
4. A lack of control of the use of governance documents. The unit responsible for developing the procedures relevant to this operation had proved unable to uncover the deficiencies in an internal revision conducted only four months prior to the incident.
5. Breaches of rules regarding safety barriers in subsea wells. The PSA notes that 'they chose to reopen a well which had been shut down due to lack of well integrity, knowing the complexity and insufficient integrity of the well' (Brattbakk et al. 2005: 24, my translation).

The PSA attributes the identified aberrations to both individuals and groups in Statoil and the drilling contractor, at multiple organizational levels and both on- and- offshore. The PSA also expressed concern that these organizational flaws were not identified before the accident. In the summary of the report, the PSA notes that they are 'critical to the fact that such extensive failures in established systems were not discovered', and that they wished to 'question why this was not detected and corrected at an earlier stage' (Brattbakk et al. 2005: 4; my translation). The investigators also noted that there was a widespread opinion in the organization that Snorre Alpha had one of the poorest HSE ratings in the Statoil organization and conclude that there was nothing that indicated that the blowout happened by chance.

The causal analysis confirms the PSAs findings, but extends and elaborates them in many ways. Of particular interest for the purpose of this study is the description of the historical, organizational and cultural context in which the deficiencies cited above are situated. In other words, the causal investigation not only investigated the immediate causes of the accident, but also included a description of safety culture at Snorre Alpha, which I shall outline in the following sections.

---

8   A risk assessment meeting for the work programme was scheduled on two different dates but was first postponed and eventually cancelled. The slot recovery operation was thus initialized without any real risk assessment.

The history of Snorre Alpha

Snorre Alpha was originally operated by a Norwegian oil company called Saga Petroleum. Snorre Alpha was the main profit generator in Saga Petroleum, a company which struggled financially and was eventually taken over by Norwegian Hydro and Statoil in 2000. According to the deal between Hydro and Statoil, Hydro operated the platform for three years before it was transferred to Statoil in 2003. Snorre had thus been operated by three different oil companies since 2000 and, consequently, had been through a period of great organizational turmoil. The people working on the platform, though, were more or less the same regardless of the changes in operator. When Snorre Alpha became a part of Statoil the strategy was to ease the installation gradually into the Statoil organization, since Snorre had already been through several changeovers. These historical aspects influenced the Snorre Alpha culture in important ways.

Safety culture at Snorre Alpha

One result of the organizational changes was that although Snorre Alpha was formally a part of the Statoil organization, it was neither socially nor culturally integrated into Statoil. The existence of such cultural boundaries meant that the Snorre Alpha organization was not part of the larger network of resources and competence in Statoil.

In addition to having been transferred between different operating companies, Snorre Alpha also experienced a change of drilling contractor as well as reorganizations in the onshore organization assigned to the Snorre field. These organizational changes coincided with the change-over from Hydro to Statoil.

Considered separately, each of these organizational changes was carried out in a way that did not pose a threat to safety. However, the fact that the changes largely happened together served to shift some of the organization's attention away from operational issues.[9]

These organizational changes were accompanied by a continuous focus on efficiency. Meeting production targets was a dominant cultural value at Snorre Alpha. The following quotation serves to illustrate this cultural trait:

> Being able to deliver the goods regarding production, budgets, plans – those are the success criteria. Doing this while also working safely – that's a challenge ... even if someone tells you that you are allowed to stop, it is hard to stop if it goes at the expense of something you have promised to deliver. (Schiefloe et al. 2005: 38; my translation)

Note that the production pressure described throughout this quotation is as much self-imposed as originating from external sources. Although the experience of a self-imposed pressure must be seen in relation to the external framework, being

---

9   This bears a strong resemblance to what was labelled 'decoy phenomena' by Turner (1978).

able to 'deliver on production' nevertheless appeared to be part of the collective identity of employees at Snorre Alpha.

The causal investigation revealed that the body of formal work requirements was seen to be so complex that it was almost impenetrable, in addition to being inconsistent and insufficiently accessible. In addition, and perhaps as a response to this, the Snorre Alpha culture was characterized by a 'can-do' attitude, where hard work and creative improvisation were central features of the dominating work practice. This materialized in an *ad hoc* form of work, where there was less room for risk assessment and working according to plans. Problems were largely tackled as they arose and there was little time and resources for long-term planning and maintenance. Schiefloe et al. (2005) make an interesting observation when pointing out that this ability to improvise and to deal with unforeseen events without having to rely on plans or procedures was essential for their ability to handle the blowout. In other words, the very same cultural traits that got them into trouble in the first place were the ones that got them out of it.

They also note that the interviewees reported some problems related to the climate for making professional objections to the actions of superiors or co-workers. This pertains to both the culture on the platform and the relation between the platform and the onshore organization, resulting in an insufficient ability to detect and correct hazardous behaviour or operations. Similar deficiencies in the organization's capacity for correction and learning are also detected in the way formal reporting systems were used. The interviews revealed that not all incidents and near misses were reported and that the organization made 'limited use of its own and others' safety experience' (Schiefloe et al. 2005: 10).

Summing up the safety situation at Snorre Alpha, the causal analysis concludes that:

> [T]he risk-tolerance on Snorre has been rather high, in the sense that relatively low safety margins regarding the verification and critical examination of potentially dangerous operations have been accepted. (Schiefloe et al. 2005: 9; my translation)

The general impression that can be derived from the two post-incident reports is the image of an organization that possesses several cultural traits that may have adverse effects on safety. In many ways, the investigators' description of the Snorre Alpha culture bears *much* resemblance to the descriptions in the investigation into the Texas City explosion, where the investigation panel found 'instances of a lack of operating discipline, toleration of serious deviations from safe operating practices and apparent complacency toward serious process safety risks' (Baker 2007: 126). This description differs quite dramatically from the picture painted through the culture survey. The next section sums up the main differences.

Before and after: diverging descriptions   The differences in the descriptions of the Snorre Apha safety culture are summed up in Table 5.3.

Table 5.3   Comparison between culture survey results and the findings of the incident investigation and causal analysis

Safety culture survey prior to the incident	Incident investigation and causal analysis
Safety highly prioritized	Safety subordinate to a dominating cultural value of 'meeting production targets'
Risk assessments carried out prior to and during work operations	Lack of risk assessments, poor understanding of risk assessment
High degree of compliance to rules and procedures, breaches sanctioned by management	Severe breaches of procedures, culture of non-compliance
Good climate for communicating safety-relevant information	Weaknesses in communication climate
Incidents and near misses reported, measures taken to prevent recurrence	Not all incidents and near misses reported, limited use of the organization's and others' safety experience
Insufficient managerial involvement	Insufficient managerial involvement

It is evident that the results of the survey and the incident investigations lead to very different conclusions about the safety culture on Snorre Alpha. In particular, there is great divergence regarding the level of compliance with procedures and the use of risk assessments. In these areas, the two sources of data give grounds for virtually opposite interpretations.

The culture survey seems to have identified some traits which were also discussed in the incident investigations. The lack of managerial involvement particularly stands out from the analysis. But this was largely conspicuous as a problem on all the other installations surveyed as well. In other words, this issue is as much a general management problem across the company as it is a unique feature of management at Snorre Alpha. Although Snorre Alpha did score somewhat lower than the other installations, these results give little grounds for concluding that there is something 'pathological' about the Snorre Alpha organization.

When faced with two conflicting descriptions of the same phenomenon, the obvious question is which of them is right. Although there will never be only one 'correct' interpretation of social phenomena, the post-incident investigations are here considered to be the most accurate as they have the benefit of hindsight regarding safety on Snorre Alpha. The experience or threat of crisis also often makes the cultures and social structures of an organization more visible (Clarke, L. B. 2006), something which also favours seeing the post-incident analyses as the most accurate.

If we accept the premise that the post-incident description is the most accurate one, the conclusion that follows is that the predictive value of the survey appears to be quite limited. This, in turn, would mean that the culture survey has largely failed in its design to assess the conditions for safety.

Possible explanations for the failure of foresight The main question underlying this study related to the ability of safety culture surveys to predict whether an organization is prone to having a major accident. The results have shown that the safety culture survey conducted prior to the gas blowout at Snorre Alpha largely failed to detect the organizational problems later identified by the incident investigations. The question that now remains is *Why didn't the survey identify more of the problems at Snorre Alpha?* Having already ruled out some possible explanations related to the validity of the survey, we can now turn to some of the other possible empirical, theoretical and methodological explanations about the divergence between the survey and the incident investigations.

'Empirical' explanations
The survey was conducted in mid-2003, while the incident investigations were conducted in the months following the incident in November 2004. Could it be that the time factor is responsible for the divergence in the two descriptions? It is possible that the safety culture at Snorre Alpha changed dramatically between the time of the survey and the time of the incident. As argued above, culture is regarded as a social construction. It is thus in a constant process of change. However, while it is theoretically possible, it is practically highly unlikely that a culture could go through such a transformation in just a year and a half that would account for the differences in the two descriptions. Culture is almost by definition a 'conservative' phenomenon. The meanings and basic assumptions that constitute the core of cultures are the very basis of the way organizational members interpret the tasks, risks and contexts of their activity. This is not something that changes rapidly without some radical changes in the tasks, risks and contexts. The Snorre Alpha organization can hardly be said to have experienced such changes in the period of time between the survey and the incident. On the contrary, the causal investigation emphasized that the Snorre Alpha culture was still characterized by the platform's history as a part of the Saga Petroleum Company. Snorre Alpha is thus characterized by a high degree of cultural *stability*, not change. In other words, one must look elsewhere in order to find the explanations for the divergence between the two views on the Snorre Alpha safety culture.

One possible explanation could be that the survey results reflect a form of complacency or feeling of being 'in control' with regards to safety, and that this created a kind of blind spot with regards to the risks involved in the work operation that led to the incident. While this may be a plausible explanation, it does not fit well with the fact that Snorre Alpha had a poor record with regard to safety performance. It is therefore hard to see how such complacency could have originated. In any case, distinguishing between 'genuinely' positive results on a culture survey and results that indicate unwarranted complacency would usually require the wisdom of hindsight. In other words, the predictive value of the survey would remain low.

Theoretical explanations

The lack of concurrence between the two descriptions can also be seen as a result of the hindsight bias[10] which is inevitable in post-factum investigations. When investigators reveal problems in hindsight, it is because they know where to look and what to look for. This raises some intriguing questions: What if we had done the broad qualitative approach *before* the incident – would such an investigation have revealed the problems that were discovered in hindsight? Would the same problems have been discovered at Snorre Alpha if the incident had *not* occurred? These questions are, of course, impossible to answer based on the data from Snorre Alpha. It could well be that the differences between the two descriptions of the Snorre culture are caused by the hindsight bias that is obviously a part of the incident investigation. Nevertheless, this does not change the fact that the culture survey did not identify the problems. The items in culture surveys are largely based on the accumulated experience of previous accident and incident investigations. This should allow for combining at least some of the analytical benefits of hindsight with the quest for foresight. Nevertheless, the symptoms were only recognized in the light of hindsight. One, rather pessimistic, interpretation of this is that proactive safety management may be impossible. It could be that the opportunities for foresight are visible only through the wisdom of hindsight.

However, this pessimistic view would be a faulty interpretation of the lessons from Snorre Alpha. As the research on high reliability organizations (HROs) has shown, it is indeed possible to learn without having to wait for accidents to happen (e.g. Weick 1987; Weick et al. 1999). The reasons for the divergence between the pre-factum and post-factum views are more likely to be related to differences in the methods and perspectives through which the Snorre Alpha organization was analysed.

Methodological/epistemological explanations

Some might argue that the data should have been analysed through more sophisticated methods, such as factor analysis, in order to provide some evidence of discrimination between scales. There will always be more sophisticated methods available for data analysis but this does not change the fact that the data itself represents a nearly consistent positive evaluation of safety culture on the installation. The focus of this study therefore remains on the way we gather data about safety culture, *not* on the various ways this data could be analysed.

---

10   The relationship between foresight and hindsight has previously been described in experimental psychology (e.g. Fischhoff, 1975), safety research (e.g. Heath, 1998; Woods 2003) and in disaster inquiries (e.g. Hidden, 1989).

Another possible methodological explanation is that the causal investigation (Schiefloe et al. 2005) has a broader analytical scope than the culture survey. The causal investigation was holistic in the sense that it analysed the way cultural, structural, technological, relational and interactional elements of the organization affected safety, and the way these elements correlated to form the context of work on Snorre Alpha. The culture survey aimed primarily at assessing values and perceptions that were assumed to constitute the cultural component of the organization. It may well be that the differences in scope contributed to the divergence between the two views on the Snorre culture. But none of this changes the fact that the culture survey did not provide an accurate description of the culture. It might, however, indicate that the questionnaire used to assess the Snorre culture is based on too narrow a model of safety culture, and that culture surveys should also aim to assess the way the cultural and other aspects of an organization interact with one another. Whether such an approach would have yielded a more valid description of safety culture on Snorre Alpha is a hypothetical question. I do not believe that any questionnaire-based study is likely to provide the depth of the causal investigation. In the next paragraphs I will elaborate the basis of this view.

I have defined culture as the frames of reference for meaning and action which encompass the skills, beliefs, basic assumptions, norms, customs and language that members of a group develop over time. Survey methods fall short of being able to assess several of the cultural elements listed in this definition. In particular, the basic assumptions that in many ways form the core of culture are impossible to reveal through survey results because they are largely tacit and taken for granted. In this respect, survey methods require respondents to be able to tell us more than they can know (Nisbett and Wilson 1977, in Rousseau 1990).

To gain information about culture requires a more interactive assessment, where the insiders, organizational members, and outsiders, researchers, of a culture engage in a process of joint inquiry to uncover cultural assumptions (Rousseau 1990). This implies taking the actor's point of view in the study of culture, instead of trying to grasp culture through the researchers' point of view by relying on *a priori* defined categories.

The comparison between the survey results and the incident investigations at Snorre Alpha illustrates the differences between the two approaches; the differences in the level of 'richness' of the two descriptions are quite striking. The results from the culture survey leave the researchers to speculate on how to interpret the findings, how the different cultural aspects – presumably – described through the survey relate to one another and how this relates to work practice.

The analysis presented here thus lends support to Guldenmund's suspicion that questionnaire data tends to reflect respondents' espoused cognitions or attitudes. In a survey about safety, it is all too obvious what will constitute the acceptable answer and the risk, therefore, is that the respondents' answers reflect the way they feel they should feel, think and act regarding safety, rather than the way they actually do feel, think and act. The causal analysis, based primarily on interviews, offers richer information as it allows for follow-up questions that are better able to generate information that goes beyond the rhetoric of safety.

The main difference between the two approaches lies in the distance or closeness to the actual work context. Culture is part of the context within which work is performed. To study safety culture is to study how this contextual factor influences and is influenced by actual work practice. This requires taking the actors' point of view and to aim at understanding how actors construct their strategies for action – why they do the things they do, in the way they do them. This requires more and different kinds of information than can be produced by questionnaires. Specifically it requires shifting the analytical focus from employees' espoused attitudes and values to their experience of their work and their social context. The goal of safety culture assessments should be to make valid descriptions of social processes and to understand why some courses of action stand out as meaningful to the actors involved. This calls for methods based on ethnographic ideals of fieldwork (cf. Hammersley and Atkinson 1995).

What is the value of survey methods for safety culture assessment?   Does the above criticism of the predictive value of culture surveys mean that survey methods are irrelevant for safety culture assessment? Although such methods certainly have their limitations I do not propose that we should abolish the use of questionnaires altogether. In some areas, survey methods can provide useful information. For instance, quantitative assessments can be useful for inter-organizational comparisons, such as the one made between Snorre Alpha and other installations.

In order to give more valid accounts of the conditions for safety, however, questionnaires should aim more at assessing the properties of the practices involved and the context of these practices. One way of doing this is to ask respondents to evaluate different courses of action to hypothetical situations. For instance, to gain information about barriers towards incident reporting, one might ask 'Suppose you were involved in an incident – Which of the following reasons may keep you from reporting this incident?' This is just an example of how one could phrase questionnaire items to assess practices and not only espoused attitudes and values. The reader is referred to Madsen (2006) for a thorough discussion of this way of assessing aspects of safety culture.

Although there is room for improvement in the survey methods used to assess safety culture, the study of culture involves a subject matter which consists of dynamic processes that cannot be fully described by means of quantification. Survey results are therefore seen as an *artefact* of culture, the meaning of which is difficult, if not impossible, to interpret without having knowledge about the cultural assumptions that lie beneath (Schein 1992). Survey methods are therefore not suited as a stand-alone approach when analysing cultural influences on safety. The use of qualitative methods can compensate for many of the weaknesses of survey methods. In studies of safety culture, survey methods are best used as part of a triangulation design where qualitative data can provide a basis for interpreting quantitative patterns. The need for triangulated research designs has previously been stressed by, among others, Grote (2008) and Guldenmund (2007).

Concluding remarks regarding the quantitative assessment of safety culture   The main conclusion of the analysis is that the culture survey at Snorre Alpha displayed low predictive value since it failed to detect most of the problems later identified by the incident investigation and causal analysis. It is argued that the lack of predictive value is due to weaknesses associated with survey methods. While survey methods can provide some information on employees' attitudes and espoused values, they cannot tell us much about dynamic social processes. Social phenomena are too complex to be adequately reflected in questionnaire items.

One may argue that these findings are neither novel nor surprising. Grote (2007: 644) stated that 'no questionnaire could ever capture the complexities of unconscious beliefs and assumptions making up an organization's culture'. But the use of surveys still continues to be the predominant strategy for assessments of safety climate which, in turn, seem to be the primary source of inferences regarding the cultural traits influencing safety in organizations and this may be one of the things that has hampered the progress of safety culture research. The aim of the study presented in this chapter has been to present empirical results that suggest a weak link between assessments of safety climate and the 'real' conditions for safety in an organization. If safety culture assessments are based on questionnaires alone, failures of foresight may be inevitable.

In spite of my scepticism about using quantitative methods to assess culture, I do not think they are useless. On the contrary, survey methods can be useful as long as they are confined to assessing phenomena that are possible to capture in the terminology of a questionnaire. In a highly enlightening essay, Denise Rousseau (1990) makes a case for multiple methods in the study of organizational culture. According to Rousseau, culture consists of many layers, structured along a continuum of the degree of accessibility of cultural 'elements' (Figure 5.1). This way of perceiving culture as analogous to the layers of an onion is strongly related to the way culture is conceptualized by Schein (1992).

This model ranges from highly observable artifacts to processes that are more or less unconscious, such as judgements about what constitutes important information. The point here is that assessing different layers requires different methods. Quantitative assessments may provide information about the outer layers, where one will find perceptions about technology, administrative routines and aspects usually associated with organizational *climate*. But to interpret the inner layers of culture, more in-depth methods are required.

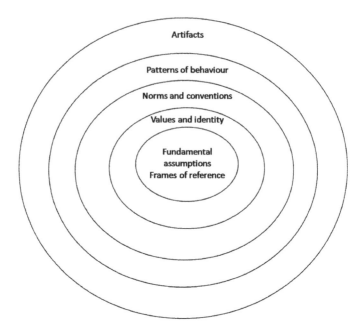

Figure 5.1   Layers of culture

Adapted from Rousseau (1990).

Ethnographically Inspired Methods

So far, I have said more about the methods that are *not* suited than I have about the alternative approaches that exist. It is now time to turn to the description of ethnographically inspired methods.

It lies beyond the scope of this book to delve into the depth and breadth of ethnographic methods of research.[11] The purpose of this section is only to describe their key characteristics.

Unlike qualitative research in general, ethnographically inspired methods are firmly rooted in the concept of culture. According to LeCompte and Schensul (1999):

> ethnography generates or builds theories of cultures – or explanations of how
> people think, believe, and behave – that are situated in local time and space.
> (LeCompte and Schensul 1999: 8)

---

11   The interested reader is referred to Hammersley and Atkinson's widely cited book *Ethnography* (Hammersley and Atkinson 1995), as well as a series of books called the *Ethnographer's Toolkit*, edited by Jean Schensul and Margaret LeCompte (e.g. Schensul and LeCompte 1999).

The origins of the concept of ethnography are to be found within social anthropology and sociology. In social anthropology, most of the classical texts are ethnographies, such as Malinowski's *Argonauts of the Western Pacific* (1922) and Margaret Mead's *Coming of Age in Samoa* (1926). Within sociology, the works of the Chicago School transferred the tenets of ethnography from their origins in the study of 'exotic' societies to Western urban environments. Whyte's *Street Corner Society* (1943) represents a landmark within sociology, partly by being an excellent empirical study, but most notably for its methodological contribution in introducing ethnographic fieldwork to sociology.

There are great differences in the way different disciplines interpret the principles of ethnography and how strictly they need to be followed. There are, however, some characteristics that seem to apply to ethnographies in general. According to LeCompte and Schensul, there are seven characteristics that mark a study as ethnographic.

The first is that ethnographically oriented assessment or research takes place in natural settings. The aim is to produce descriptions and understanding of events as they occur in their natural context. This means that the researcher should have gained first-hand knowledge about the phenomena under study. Those looking to understand safety culture should become familiar with the contexts where work is performed, decisions made and safety issues discussed. Spending some time in these areas, observing what happens, listening to what is being said and asking questions to understand the field under study, are prerequisites for the creation of a thorough description of safety culture.

The second is that ethnographically oriented researchers strive to become intimately involved with members of the community under study and engage in face-to-face interaction with them. This involves building trust between the researcher and the members of the community, something that usually requires a great deal of time and persistent effort. In effect, ethnographically inspired methods dispense with the traditional distance between researcher and the research subjects, making the researcher the primary tool of measurement. As I discuss further towards the end of this chapter, removing this distance creates both opportunities and challenges for making accurate descriptions of cultural traits and processes.

Thirdly, ethnographic assessments place emphasis on the perspectives and interpretations of the participants in the research. The goal of ethnographically inspired research is to provide accurate descriptions of the way informants perceive and make sense of the world around them. This involves an attempt to 'imagine the other', to see things from the participant's view, and this, perhaps, is the core principle of ethnographic research. Adherence to this principle requires trust between the researcher and the participant; trust that is vital to ensure access to information about the participants' view and interpretation of the world around them.

Fourth, LeCompte and Schensul (1999:15) stress that ethnographic methods use 'inductive, interactive and recursive processes to build theories to explain the behavior and beliefs under study'. This means that the interpretations made about cultural processes in the field under study are produced through dialectic between data and hypotheses. This view on the relationship between induction and the construction of models or theories is closely related to Glaser and Strauss' notion of 'grounded theory'. This means that cultural descriptions will be based both on inductive inferences from data and a more deductive process where the inductive inferences are related to existing theoretical knowledge.

The fifth characteristic is that ethnographically oriented methods use multiple sources of data (Hammersley and Atkinson 1995; LeCompte and Schensul 1999). Any form of data that may shed light on some problem may be included in ethnographic studies (ibid.). Accident/incident data, data from questionnaires, observation, structured or unstructured interviews, learning histories, narratives or search conferences, all may be included in the ethnographer's toolkit. This being said, the case study from Snorre Alpha indicates that quantitative methods should be reserved to gather information that does not require interactive probing. While survey methods may not be well-suited for providing accurate information about cultural process, as defined in this book, they can be used to provide contextual information,e.g. work environment, risk perception, etc.

The sixth is that knowledge about social phenomena is always seen in relation to the context in which they occur. Here the term context refers to the various elements, e.g. technology, history or political factors that constitute the framework conditions for the creation and re-creation of culture.

And lastly, it is the hallmark of ethnographically inspired assessments that they are informed by the concept of culture. Ethnographically oriented assessments are concerned with providing a description of why people do things, say them or believe in them.

Since it is the primary methodological approach to the study of culture in mainstream social science, the inclusion of ethnographically inspired methods in the toolkit of safety research is probably long overdue. I should stress though that ethnographic methods are not the answer to all methodological challenges related to safety assessment. Moreover, the very nature of safety issues may impose some limitations on applying ethnographic ideals within safety research. One example; studies of safety will very often involve examining phenomena or processes that take place in high-risk settings not always accessible to outsiders. In addition, the principle of spending months or even years in the context where the study takes place is generally difficult to combine with the researcher's family obligations or other work commitments. Both of these issues may serve as effective barriers towards living up to the tenets of full-fledged ethnography. Allowing for these practical problems, one should aim at the greatest level of depth that is practical and possible to combine with leading a normal life. The difficulties of meeting all the demands of ethnography is the reason why I prefer to speak of ethnographically *inspired* methods, not ethnographic methods as such. Although it is certainly desirable to spend a lot of time in the natural context of the phenomena under study, the main strength of an ethnographic approach lies less in the amount of time spent in the field than in its basic stance towards the actors in the field where the study takes place. The aim of the data collection is to provide descriptions of *their* perception and interpretations of issues of risk and safety and this stance provides a very different basis for the design and carrying out of an assessment, when compared to more 'distanced' methods.

In the next chapter, I shall illustrate how an ethnographically inspired approach can be applied within safety research.

Chapter 6

An Empirical Case Study – Safety Culture on Offshore Supply Vessels[1]

The methodological arguments raised in the previous chapter are by no means novel. Calls for the use of qualitative approaches and triangulation between qualitative and quantitative methods have been voiced by several scholars (Rousseau 1990; Glendon and Stanton 2000; Guldenmund 2007; Grote 2008). So far, with notable exceptions, such as Richter and Koch 2004; Farrington-Darby et al. 2005; Nævestad 2008, there have been few attempts to adopt this methodological standpoint in empirical investigation . This chapter is an attempt to remedy some of this shortcoming and to show how the methodologies considered in previous chapters can be utilized to produce empirical research.

This chapter examines the relationship between occupational culture and safety on offshore supply vessels in the Norwegian petroleum industry. The term occupational safety refers to cultural traits shared by those who follow an occupation, in this case seamen, and that transcend organizational boundaries (Van Maanen and Barley 1984). Some of the cultural traits and processes described are, however, also related to the logistics chain of which the vessels form part. Following the discussion in Chapter 1, we can say that what constitutes the concepts of both organizational and occupational culture is very similar, relating as they do to the tenets of how work should be performed and how the members of a community see themselves and outsiders. Here the term safety refers primarily to occupational injuries.

The analysis presented in this chapter combines qualitative and quantitative methods in a way that is consistent with the ethnographically inspired methods described in Chapter 5. The aim of the study is to give a general depiction of the professional culture aboard the supply vessels and the way this culture may influence safety. I shall show through this analysis that in order to understand the way culture influences safety it is not sufficient to address cultural issues alone. Rather, one should employ a more holistic perspective, emphasizing the interplay between cultural, structural and interactional aspects of work.

---

1   This chapter is partly based on an article published in *Safety Science* – see Antonsen (forthcoming).

The Organizational Context of Supply Vessels

The oil and gas industry is, without doubt, the key industry in Norway, representing 15 per cent of GDP. The offshore supply vessels are an important part of the logistics chain involved in the production of oil and gas on the Norwegian continental shelf. Almost all goods or equipment needed by the installations is transported by supply vessels. To a great extent the activity in the logistics chain is controlled and coordinated by the oil company, but several organizations are involved including the ship-owner's offices, the ships' crews, supply bases, the oil company with its different departments and the crew on the oil and gas installations. This chain of actors and activities constitutes the organizational environment surrounding an offshore supply vessel.

Aboard a Norwegian supply vessel one will usually find a crew consisting of 10–15 seamen, working rotating six hour watches in a 28 day shift. The vessels usually visit port about three times a week to load fresh and unload return cargo from the installations. After 28 days, the crew has a period of four weeks off, during which the crew scatters until it musters for a new shift onboard the vessel. All the crews studied consisted of ethnic Norwegians only.

The crew typically consists of a captain, a first and a second officer, a chief engineer with two or three engineers, a cook, at least four sailors and sometimes an electrician. At one time the social relations onboard ships were marked by a highly visible status hierarchy and an authoritarian model of management. The captain was 'next to God' and it was not common for officers to socialize with the rest of the crew (Aubert and Arner 1962) but this kind of situation is rarely found in Norwegian crews today. In one way, of course, this reflects general changes in society towards greater democracy at work, but it is probably also because crew size has decreased dramatically in recent years. Today, the crews of most supply vessel crews are characterized by a sense of – literally – being in the same boat, although obviously still organized through hierarchical and functional division.

In the early research on shipping, ships were often described as 'total institutions' (Arner 1961; Aubert and Arner 1962; Schiefloe 1977), a concept borrowed from Goffman's (1961) study of psychiatric hospitals. A total institution is characterized by a suspension of the division between different spheres of life, e.g. work and leisure time (ibid.). A prison is perhaps the most common example of a total institution. Similarly to prisoners, seamen are confined to a limited physical area where they eat, sleep, work, and socialize with the same people for long stretches of time. The prison-like aspect of ships is illustrated by the afore-mentioned expression of 'being in the same boat', referring to a sense of being stuck in the same situation as several other individuals and thereby having something in common. In total institutions individuals also possess a low repertoire of roles compared to life outside them. The role of captain on a ship, for instance, affects all aspects of interaction between captain and crew members irrespective of whether it happens on or off duty. There is virtually no other role that the individual with the role of captain can enter into when interacting with his crew.

In spite of all the advances in communication technology and the fact that the supply vessels call into port several times a week, the research of the 1960s still provides valid accounts of what it means to be a seaman. Although crew members now have far more contact with the rest of society, they are still isolated from family, friends and the rhythm of ordinary life. Lamvik (2002: 66) captures the essence of this aspect of the seafaring experience by describing life on the sea as a 'state of exception'. To be a seaman means being a part of an alternative society which has no counterpart elsewhere and although Lamvik's study is of seamen in the merchant fleet this is a fitting description of life on a supply vessel as well.

Because they are miniature societies ships make unique cases for studying social phenomena like culture, and the way culture may influence safety. Crew members interacting very closely with the same people over long periods of time constitute fertile ground for the creation of distinct culture traits. At the same time, its confined nature provides the researcher with an opportunity to study a type of social system less susceptible than other organizations to outside influence.

Injury Statistics

The injuries occurring on supply vessels are mainly occupational in the sense that most accidents involve only one or a few employees. The most risky operations on a supply vessel require cooperation and coordination between several organizational units, which means therefore that some accidents will have the characteristics of organizational accidents. When navigating the vessel close to the platforms there is always the risk of colliding with them. Depending on the design of the platform, its leg may be punctured and the subsequent water intrusion could cause the platform to capsize. Thus, the activity on a supply vessel also involves some potential for major accidents.

Table 6.1 gives an overview of the frequency of deaths, injuries and incidents requiring medical treatment, collisions and the aggregated injury rates of all service vessels employed by the oil company.[2]

The most serious injuries tend to happen when the vessels are loading and unloading alongside the platforms and particularly so in bad weather. Wave heights of six or seven metres create a challenging work situation for both the ship's navigators and the sailors working on the cargo deck, and, of course, such weather increases the chances of the vessel colliding with the platform. Waves washing over the cargo deck may cause containers to shift and hit the crewmen, in extreme cases washing them overboard.

---

2   The number of incidents and injury rates include incidents from stand-by vessels and anchor-handling vessels in addition to the supply vessels. The number of collisions includes only supply vessels.

Table 6.1   **Overview of incident (first aid incidents excluded) frequency, collision frequency and incidents per million work hours**

Year	Number of incidents	Injury rate	Number of collisions between vessel and platform
2000	17	5,0	12
2001	26	6,9	1
2002	15	5,2	1
2003	12	6,6	1
2004	10	6,1	0
2005	5	2,6	2
2006	14	7,9	0
2007*	19	7,6	1

* Not fully comparable to previous years due to significant organizational changes.

Research Methodology

The triangulation approach used here combines the wide-angled view of survey design with the depth of qualitative methods like interviews and search conferences. The quantitative data was collected in the form of a self-completion questionnaire, consisting of 95 items. It aimed to seek the opinions of the seamen on the following topics:

- Incident reporting;
- Company managers', captains' and seamen's prioritization of safety;
- Work procedures;
- Work situation; work stress;
- Competence and training;
- Communication and cooperation;
- Management;
- Lines of responsibility;
- Perceptions of seamanship.

The questionnaire was developed on the basis of a review of existing questionnaires and the survey aimed to give a broad, introductory analysis of the relationship between organizational factors and safety on the supply vessels. The questionnaire was given to the crews of all supply vessels contracted by the oil company in question. 258 (56.7 per cent) completed questionnaires were returned. The survey data was processed through analysis of variance (one-way ANOVA) and simple frequency distributions.

In addition to the survey, data was also collected through in-depth interviews, observation of work and participation in search conferences. 22 interviews were conducted with a total of 29 informants;19 individual and three group interviews. The study focused on two shipping companies contracted to the same oil company. Three informants were HSE personnel at the shipping companies, while four were safety managers in the oil company. These informants were included to obtain some information about the larger system of which the supply vessels are part. The remaining 22 informants were working on supply vessels from the two shipping companies. The interviews with these informants represent the primary source of data.

Some of the interviews with crew members were conducted while the vessels were in port but most of them were conducted while at sea while the researchers spent two to three days at a time onboard the vessels. In addition to conducting interviews, being onboard the ship for several days provided researchers with opportunities to have informal conversations with crew members and observe the way work was carried out. Collecting data on board the vessels has a couple of advantages. First, it is much easier to gain the informants' trust when they are on their home ground and when you have the opportunity to interact informally with them over a longer period of time. Second, collecting data in the context of work allowed for rich examples that helped to interpret it.

The interviews were recorded and later transcribed. The transcribed texts were then categorized into different themes that stood out as important. These themes included both deductive categories, in the sense that they were derived from theory, and inductive categories that originated in issues that the informants brought up (Miles and Huberman 1994). These categorizations then provided the basis for searching for common patterns in the interviews. This is a process first of decontextualization in the sense that it involves breaking down the meaning content of the interviews into separate units. In order to interpret the data a process of recontextualization is necessary, i.e. a return to the big picture, asking how the identified themes and patterns relate to each other (Thagaard 2003). The product of this analysis was the identification of some common categories through which the seamen described themselves and others, as well as conventions for work performance, interaction and communication.

In addition to these more traditional methods for data collection, I also attended a total of seven so-called search conferences involving supply vessel crews. The basic idea behind search conferences is to create participative arenas directed at both describing organizational problems and searching for strategies to solve them (Levin and Klev 2002), in this case safety related issues. Engaging in the process of 'searching' (Emery 1999) meant that the seamen participating in these conferences were in reflexive mode, which made the conferences suitable arenas for learning more about the way the seamen thought about safety, and how their views resonated with the safety policies of the oil company. In addition, the conferences provided good opportunities for validating my interpretations and inferences based on the interview data. The combination of interviews/ observation onboard the vessels and the learning from search conferences in my view serves to strengthen the data since it encompasses both the way the informants reflect *in* action, and the way they reflect *on* action, to use the words of Schön (1987).

Culture and Ssafety on Offshore Supply Vessels

The purpose of this study was to present an empirical analysis of the relationship between occupational culture and safety on Norwegian supply vessels. Since in my view the description of the cultural traits of an organization or an occupation logically needs to come before any discussion of how these traits influence safety, these two questions will be dealt with separately.

Cultural Traits of Work Communities on the Vessels

The supply vessels are 24-hour societies, where the crew members perceive themselves to be 'out of phase' with the rest of the world (Zerubavel 1981: 67, cited in Lamvik 2002). This strongly influences the conventions for language, behaviour and social interaction. The language conventions on the supply vessels typically allow quite rough language, especially among the older men. This is not surprising considering that the crews are almost always male, and that swearing is often associated with traditional masculine values such as independence, spontaneity and strength (Ljung 1987).

Behaviour and interaction between crew members appear highly codified. For instance, there were very clear expectations that the crew members should spend some leisure time in the mess room. They rarely retreated to their respective cabins until they went to bed, and if they did they almost always kept the door to the cabin open. An open door signals that one can be disturbed, while a closed door is a signal that one wishes to be left alone. In fact, when asked how they dealt with interpersonal conflicts and differences, several of the seamen interviewed mentioned retreating to one's cabin and shutting the door as the way *not* to deal with such issues. In many respects, retreats to privacy were interpreted as intent to 'resign' from the community and as such a breach of the conventions for proper conduct. The existence of this and similar conventions tells us two things. First, that a ship is a system with a high degree of social control. There are many spoken and unspoken rules for what is considered the proper way to relate to each other. Second, it shows that the existence of a collective or community is highly valued by the crew members. The importance of having a good atmosphere and tone between crew members was a striking feature of the way the seamen described their own situation. Despite the specialized division of labour and the hierarchical organization, there was still a sense of solidarity and cohesion that referred to the vessel crew as a *unit*. In an interview with a first officer, the informant emphasized that relational skills were a central attribute of being a good seaman:

> Sometimes you have to apologize, and sometimes you get an apology. To adapt to others, learn from your mistakes and admit your own mistakes, that's [important], because you have no place to hide here! You have to cooperate.

The importance of maintaining social relationships and community was also evident in the survey results. The survey included an item where the respondents were asked to state five words that they felt characterized the notion of 'good seamanship'. This item was included to assess aspects of the seamen's occupational identity. This collective identity, in turn, is an important part of organizational culture. Table 6.2 presents a rough categorization of the words used to answer this item. It shows that the ability to maintain social relations and community was the most frequently given characteristic of good seamanship.

The emphasis on community and cooperation should not be taken as an indication that there are no conflicts onboard the vessels. On the contrary, personal and professional conflict is a common part of life onboard. This is probably inevitable when a group of people are 'stuck' on the same ship for 28 days. Nevertheless, there seem to be some social conventions for how conflict should be dealt with in order to prevent it from becoming too destructive for the onboard community. For instance, if any of the officers on the bridge spot a mistake being made on the cargo deck, it is not considered appropriate to communicate this over the radio, since all crew members involved in an operation would be on the same radio frequency. In order to avoid such a public correction this kind of thing would be commonly dealt with in a more informal arena, for example at the next meal. Therefore, the seamen's emphasis on support, cooperation and a sense of community has two sides. It refers just as much to the ability to *deal* with conflict as to avoid it.

Table 6.2   Overview of the respondents' characteristics of the notion of 'good seamanship'. The number of respondents stating each characteristic in parenthesis

Crew relations (153)	Support, trust and cooperation skills (96)
	Strong sense of community, ability to uphold social relationships (57)
General ideals for performance (151)	General ability to work safely (108)
	High quality on the work done (43)
Individual properties (85)	Independent, resolute (27)
	Responsible, reliable (58)
Competence (80)	Practical sailing experience (54)
	Attentive, careful (26)

The second most frequent category pertains to the standards for work performance, where the ability to work safely and provide high quality services is highly valued. This must be seen in relation to the third largest category of descriptions, relating to the ability to make wise and cautious decisions. The ability to make sound judgments about work related risks falls within this category. The individual properties describe traditional masculine values of autonomy, independence and decisiveness. Being a good seaman seems to have much to do with an ability to work on one's own initiative, i.e. without having someone telling you what to do, and to deal with unforeseen events as they arise.

Other characteristics that stand out as central parts of the notion of seamanship are the ability to 'get the job done', being able to take care of novices, and in particular *having sufficient practical work experience*. This last seems to be a criterion that pervades most other characteristics of good seamanship. In her study of seamen's work identity, Knudsen (2005) notes that the acquisition of sailing experience is an important rite of passage in ship culture. Without possessing this experience it is very hard to speak of life and work at sea with any authority. This feature was also present in the occupational contexts studied here. The seamen perceive of themselves as experts in their field and the legitimacy of this expertise is based on the acquisition of practical sailing experience. As I shall discuss in detail below, this is a criterion of legitimacy that poses a challenge to safety management in the shipping industry.

In addition to analysing the words the seamen use to describe the model of their work, it is also worth considering the words that are *not* used, and which consequently cannot be regarded as an explicit part of the notion of good seamanship. Only 22 of the 258 respondents mentioned working according to formal procedures as a characteristic of seamanship.

There is very little variation in the words used by the different work groups to describe the concept of seamanship. This can be taken as an indication that the identity and work ideals expressed through the notion of seamanship seem to cover all seamen, irrespective of their position in the ship's social system. Being a seaman and having good seamanship seem to be categories referring to a set of core values about life at sea.

To sum up, the presumption conveyed through the notion of seamanship is of autonomous men – the premise is predominately masculine – who are able to make safe decisions based on their own professional skill and judgment. The seamen take great pride in their profession and their ability to provide reliable, high-quality services. This amounts to a form of 'can-do' culture, which is at the same time both service oriented and self-driven. There is a high level of willingness to work hard but a great deal of reluctance towards being told *how* this work should be done, especially by people who are not themselves seamen. Furthermore, the ability to meet the conditions of good seamanship is not something learned through formal education but largely based on first-hand sailing experience.

There is a great deal of professional pride connected to the notion of good seamanship. Some men express a great deal of frustration that the traditional characteristics of good seamanship are in danger of being lost:

> You know, good seamanship, it is tragic, it is about to disappear completely. That expression, 'good seamanship', it doesn't exist anymore, because everything that is to be done, has to be written on a list. You are not supposed to use good seamanship and common sense, you are supposed to use check lists, procedures and maintenance lists. That's what it's all about. And I know this is a source of great annoyance to the guys on the deck.

This quotation illustrates that the traditional ethos of seamanship is perceived to be under attack by attempts to regulate work by formal regulations and written documents.

The core values of seamanship are not just what give professional identity to the seamen. They also constitute important lines of demarcation between 'us' and 'the others' (Bye et al. 2008). Two groups in particular were contrasted with the seamen's perception of their own occupational and professional integrity. One of them was the 'landlubbers' at the ship-owners office and in the onshore organization of the oil company. Since few of these people have worked at sea, many crew members expressed a certain scepticism towards the competence of these office workers, or 'forest workers' as they are sometimes labeled to highlight their lack of sailing experience, when it comes to making decisions concerning the vessels. And within this attitude the seamen maintained that administration and paper work lie outside their conception of a seaman's duties:

> You know, us seamen, we're not so keen on filling out a whole lot of forms. That's not where our job lies. Our job is to sail a boat, we're not office workers whose job is to fill out a bunch of forms.

The other group was the workers on the offshore platforms. The platforms are the indisputable centre of activity in the logistics chain since they produce the oil and gas and generate the income. To hold a central position in the division of labour is an important source of power (Pfeffer 1981). Compared with the seamen on the supply vessels, the platform workers are in a privileged position when it comes to power and status. The platform workers work shorter shifts, make more money and have better working arrangements than the seamen. In addition, the platform workers are usually employed by the oil company while the seamen are 'mere contractors'. These differences in power and status are also symbolically expressed in the relationship between platform and vessel. The seamen especially express indignation over the soiling of the vessels by dirty spill water or drilling fluids. They see this as an expression of disrespect – that the platform workers see the vessels as dumpsites. There are also stories about platform workers throwing trash and even urinating down on the vessels (Bye et al. 2008). Such narratives may contain important information about status and identity (e.g. Czarniawska 1997; Jovchelovitch and Bauer 2000; Amundsen 2009). The common theme in all the narratives containing comparisons between seamen and platform workers is that the seamen perceive themselves being at the bottom of the status hierarchy of the logistics chain and that the supply vessels are treated as a mere service function that exists to satisfy the whims of the platforms.[3]

Such narratives of course may not give accurate accounts of the events . They may be both exaggerated and misleading. For instance, the spills of water or chemicals that soil the vessels are most often unintentional. Irrespective of their objective truth, however, the *symbolic* message for the seamen remains the same. The organizational folklore encapsulates a deal of informal knowledge about the character of the work and the way the group relates to other groups. As such, this folklore expresses the explicit and tacit contents of a group's identity. In these aspects of occupational identity the vessel crews are positioned as the 'underdogs' in relation to the oil and gas installations. As will be discussed in further detail below, these status differences may influence decisions as to what constitutes safe working conditions.

A note on ambiguity   When researchers describe cultural traits, these descriptions are always simplifications of complex social phenomena. Since simplification necessarily involves toning down the nuances of the data, I need to say a little about things that lie in the background of the above picture of cultural traits.

First, there were some variations between older and younger seamen, where the older exhibited a closer relationship to the ideal character of seamanship. This is not surprising, since the notion of seamanship is firmly rooted in maritime history and tradition. Also, there were variations in social conventions and the tone of communication between different ships. Because culture is constructed through social interaction it should not surprise anyone that there will be cultural variation between groups that are isolated from each other. In other words, the cultural traits described above are to some extent ambiguous in the sense that they are not interpreted equally by all members of the organizations and that there is no unanimous definition of *the* organizational culture on the ships.

But there are quite striking resemblances between the ways most of the seamen view and conceptualize their own work situation and the ideal standards for work performance, which legitimizes singling out the cultural traits described above, and viewing them as representative accounts of the seamen's occupational identity. The cultural traits described here are, however, to be viewed as *ideal types*, generalizations that summarize rich and complex phenomena into analytical

---

3   The oil company in question has taken several measures to improve the status and working conditions of seamen on the service vessels. For instance, there have been efforts to increase the awareness on the platforms towards the problem of soiling supply vessels. While the problem is nowhere near eliminated, there seems to have been some progress in reducing the occurrence of emissions on to the vessels (Fenstad et al., 2007).

constructs. There will be variations as to how much a group resembles the ideal type, but all the vessels analysed here display some level of these cultural traits. Moreover, the descriptions of ship culture made in the above analysis are very much consistent with previous research on ship culture (e.g. Lamvik 2002; Knudsen 2005), something which strengthens the validity of the generalization.

The Relationship Between Cultural Traits and Safety

Ship Culture and Communication Climate

The sense of solidarity and community – 'being in the same boat' – that stood out as a cultural trait of the ship crews, has several effects generally regarded as positive in terms of safety. The crew relations seem to foster a climate for care and open communication. Several items in the survey illustrate that this communication climate also encompassed safety issues. Table 6.3 shows the mean and frequency distributions on selected items regarding communication climate.

The frequency distributions in Table 6.3 show that the seamen are highly positive towards the sharing and discussion of safety related information, both between co-workers and between crew and officers. Also, as illustrated by item 5, the informal correction of unsafe behaviour seems to be socially accepted on board the vessels. These features, all related to the vertical and horizontal flow of safety related information, constitute the key prerequisite for safety in Westrum's (1993) ideal type of a safe organization, the generative organization.

Table 6.3   Questionnaire items related to communication climate among crew members. N, mean and frequency distribution (percentages)

Questionnaire Item	N	1 (%)	2 (%)	3 (%)	4 (%)	5 (%)
1. My supervisor appreciates the employees coming forward with safety-related issues	252	1	2	7	35	56
2. I'd rather not discuss safety related issues with my supervisor	254	68	17	6	7	3
3. Do you discuss safety related issues with your co-workers?	255	1	3	16	46	35
4. The working environment on board is characterized by openness and dialogue	256	1	7	8	34	50
5. On board my ship, it is common to speak up if one sees colleagues working in an unsafe way	256	5	3	7	37	48

1= strongly disagree, 2= disagree, 3= somewhat disagree, 4= somewhat agree, 5= agree, 6= strongly agree.

The tightness of the on board community may have its downside. Although such tightness tends usually to be associated with open communication some informants see the strong social control onboard as conservative and conflict-averse. Personal relationships between the crew members are characterized by a very narrow status repertoire since there is very little separation between personal and professional relations on board and this may make it difficult to deal with professional matters in a detached way. There is some relationship here to the concept of groupthink. Efforts to maintain the integrity of the community may involve minimizing conflicts in order to reach consensus and in the process shut out information and ideas that deviate from the prevailing definition of reality (Janis 1982). So a tight, homogenous culture is not necessarily positive for safety.

'Putting it in writing' – Between seamanship and procedures   The cultural view of what it means to be a competent seaman, conveyed through the notion of 'good seamanship', expresses attitudes toward behaviour and work that may influence safety in a number of ways. Firstly, the view provides frames of reference through which risk and safety is evaluated and, in particular, an image of what it means to work safely. According to the notion of good seamanship, the ability to work safely is based on the prudence that comes with practical sailing experience. This means following intuition, working in line with professional judgments and the peculiar circumstances of each situation, and improvising when necessary. There appears to be a great deal of inconsistency between these *informal* models for work performance and the *formal* models conveyed by safety procedures. Formal work requirements seek to standardize work processes and reduce the necessity for improvisation. This inconsistency could be conceptualized as friction between the structural and cultural elements of the ship organization.

One way this friction is given expression is through the seamen's frustration at being 'forced' to work by formal rules:

> Suddenly you are supposed to put words on something you have done your whole life. You know how to do this – why on earth do you have to have a checklist about it? Why do you need a procedure on it? Why does it have to be in writing? We know how to do this! And then it's like 'don't they trust that I know how to do my job?'

The data is full of similar statements, illustrating that attempts to govern work by formal rules are often interpreted as a negation of the seamen's professional expertise. Moreover, formal procedures usually have their origins *outside* the community of seamen, in onshore organizations like regulatory authorities, oil companies or the ship-owner's office. This, trivial as it may seem, has a strong influence on the way the seamen see attempts at formal safety management. The fact that they do not see such procedures as based on the practical knowledge possessed by competent seamen but rather on the theoretical knowledge of some 'office worker', for them undermines the legitimacy of formal procedures.

The result is that the seamen may see formal safety procedures as having little relevance to their work and therefore often have a somewhat relaxed attitude towards them. The survey results provide an example of this. The seamen were asked the question: 'if you fail to comply with a procedure, what could be the reasons for this?' The respondents could choose up to three alternatives. The frequency distribution of this item is portrayed in Figure 6.1. It is interesting to note that only 17 per cent of the sample of 258 persons answered that they never violated procedures.

Reasons given for not working in accordance with procedures show that going by the book is sometimes perceived to take longer, require a great deal of specialized knowledge, and that the procedures do not work as they are intended to. This last reason should be seen in relation to the above-mentioned criterion of legitimacy. The seamen sometimes seem to question whether formal procedures serve to increase safety, something brought out in an interview with a first officer:

> Some of the procedures we take very seriously. But running around telling the sailors how to perform every task, what equipment they are to use, how they are supposed to stand when using an angle grinder – I cannot relate to that! . . . On many occasions, I have to say, I have a hard time finding the reasons for my doing this. I cannot see how this increases the safety of the person performing the task.

I must stress that this discussion of the way the seamen related to procedures should not be taken to imply that they do not put a high priority on safety. The seamen engage fully in safety issues and, as indicated above, the ability to avoid injuries is a part of what is regarded as good seamanship. It should also be noted that most of the seamen interviewed accept that some formal procedures are necessary,

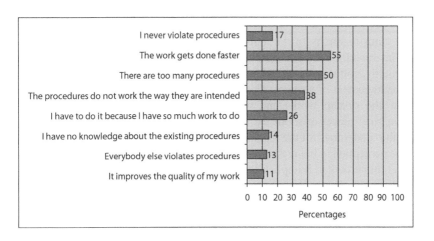

Figure 6.1   Reasons for violating procedures (percentages)

especially those pertaining to work operations involving high risk and/or that are performed on rare occasions. The primary target of criticism of formal procedures seems to be the logic behind the checklist (Gherardi and Nicolini 2000: 13) which vessel crews see as imposing a technical and managerial rationality onto the nitty-gritty details of everyday practice. The seamen put a very high value on safety but nevertheless disagree on the best way to achieve it, a good illustration of the way different frames of reference can give quite divergent views on what is considered dangerous or safe.

The point of the above discussion is thus *not* to show that the seamen's resistance towards working by the book is a threat to safety and that all would be well if they would only get the 'right culture' and comply with procedures. The point is to show that the cultural traits of an organization influence the way the organizational structures are perceived and that an efficient safety management system is created in the interplay between aspects of culture and aspects of structure. Safety is, to a great extent, a matter of maintaining control over variations in the performance of a system or work organization (Hollnagel and Woods 2006). This, in turn, requires a high degree of knowledge about the system actually works. Any gap between procedures and practice may mean that managers have very little information about the way things are really done at the sharp end and this is where safety problems can arise.

The match between formal and informal aspects of work is therefore a key issue for the study of safety culture, as has previously been argued by Hale (2000). On the supply vessels I have studied here there seems to be a rather low degree of match between culture and structure, resulting in a loose coupling between procedures and practice. However, there are signs of change in this respect. The oil company in question has acknowledged this problem and started to look at other ways of developing and implementing procedures (cf. Antonsen et al. 2007).

Safety consequences of the asymmetrical relationship between platform and vessel As I wrote earlier, the seamen perceive themselves to be at the bottom of the status hierarchy of the logistics chain. This is given symbolic expression in the interpretation the seamen put on those instances where platforms, probably accidentally, soil the vessels with spill water etc. and in other stories suggesting the disparagement of seamen. These narratives express the seamen's conception of how the system works and what is expected of the supply vessels. The asymmetrical power relations expressed through such stories create a feeling of being the 'underdogs' of the logistics chain (Bye et al. 2003).

The perception of being the weaker party seems to influence safety in a number of ways. In particular, it implies adopting what may be described as a service role towards the platforms. When weather conditions are marginal this role seems to exercise influence on the decision as to whether it is safe to go ahead with the loading and unloading of goods alongside the platform. Several of the interviewees tell of situations where they have agreed to a platform's request to start working despite having doubts as to whether it is safe enough. I must stress that this is not necessarily due to explicit pressure from the representatives of the platforms. All informants stress that there have been considerable improvements regarding direct pressure from installations. Nevertheless, the interviewees will say they felt pressure or expectation to go to considerable lengths to satisfy the needs of the oil and gas installations, who are at the center of the value chain in the petroleum industry. In poor weather, pushing the limits of safe operations no doubt increases the risk of the vessels crashing into the platform legs. There have also been some serious accidents where the men on the vessel deck have been badly injured by heavy sea breaking in over the deck and causing containers to start sliding. In other words, the asymmetrical power relations between vessel and platform seem to influence decisions about when working conditions are considered safe enough.

Asymmetrical power relations have already been identified as contributing to major accidents. The *Challenger* case discussed in Chapter 4 is a prime example of how asymmetrical power relationships may influence safety. A more recent example is the accident of the anchor-handling vessel Bourbon Dolphin, which capsized and sank during a rig move west of Shetland in 2007. The investigation of this accident suggested that the captain and first officer although concerned about the risks involved in some aspects of the approach allegedly suggested by the towmaster, nevertheless agreed to follow the towmaster's approach (JPD 2008).[4]

Of course, there are more to these accidents than the asymmetries of power, but they nevertheless indicate that the role of such asymmetries in safety-critical decisions should not be underestimated.

Concluding Remarks and Implications

As should be clear from the previous sections, the occupational culture of seamen displays several traits that influence safety. In addition to the empirical findings, the analysis also has some theoretical and methodical implications.

First, it shows the advantages of splitting the analysis of safety culture into two separate analytical steps. Describing the cultural characteristics of the group in question is the primary research task, only after this is done can we discuss the way the identified cultural traits influence safety. The advantages of such an approach lie in the richness of the data generated from the analysis. Placing organizational culture at the center of the analysis directs the researcher's attention towards the general aspects of informal organization and the way cultural traits are embedded

---

4   There is disagreement over whether the towmaster actually suggested that the Bourbon Dolphin should lower the inner towing pin holding the anchor chain. The second officer onboard the Bourbon Dolphin stated in his testimony that the towmaster did make this suggestion. The captain and first officer were reportedly anxious over the suggestion, but nevertheless agreed to lower it. This operation was probably the triggering cause of the accident. The towmaster denies having made any suggestion to lower the inner towing pin. Both the captain and first officer were killed in the accident.

in a social and organizational context. This allows for 'thicker' descriptions (Geertz 1973) of cultural traits. Compared to this approach, a traditional safety culture approach tends to 'jump to conclusions' since it does not take the time to do a thorough description of the culture in question. The latter approach may tell you that the employees have a lax attitude towards procedures, or that they sometimes continue to work in unsafe weather conditions, but it will not tell you *why*. Knowing why, that is why certain types of behaviour and decisions stand out as the right things to do, is, in turn, crucial in order to know what to do about it.

Second, the analysis indicates that occupational identity should be considered a key part of the concept of safety culture, as stressed by Gherardi and Nicolini (2000). Where a group draws the line between what they see as insiders and outsiders, how they think of themselves and of their work, are all important cultural dimensions that are likely to influence safety. Whether they in fact influence safety in concrete organizations is an empirical question. It is also instructive that the way the seamen differentiate between themselves and others constitutes a commonly recognized division between the two fundamentally different forms of expertise; with practical craftsmanship on one side and theoretical, generalized knowledge on the other. The extent to which these two forms of expertise relate to and respect one another is an important part of the relationship between professional cultures and rule-based organizational structures.

Third, the analysis illustrates that cultural traits interplay closely with other aspects of organization such as formal structures and the interaction between different groups and that the degree of 'match' between the different properties of the organization is crucial for the ability to manage safety. We could see that on the supply vessels there were incompatibilities between culture and structure that led to resistance towards formalized safety procedures. We also observed the role of power differences in determining what is 'safe enough', suggesting that the interaction between different groups is important for understanding cultural influences on safety. The interplay between the cultural, structural and interactional aspects of organizing calls for a more holistic approach to the study of safety in general and safety culture in particular. Such an approach should include a more political perspective on organizations and question how asymmetries of power influence culture and safety.

It should also examine issues of organizational symbolism, largely neglected by previous research on safety culture. After all, it is in the nature of cultural phenomena that they are largely taken for granted by the members of a culture. In other words, as outsiders, we cannot expect the members of an organization to be able to present a reflexive picture of how they see their own culture.

In order to study culture we have to be sensitive towards the objects, narratives and behaviour that seem to be invested with meanings for the members of a given community. This requires an ethnographically inspired approach to the study of safety culture.

Chapter 7

Improving Safety Through a Cultural Approach – Limitations, Constraints and Possibilities

Safety research is often applied research in the sense that it is intended to have some kind of practical utility, the knowledge produced should provide the basis for improvements. Compared with the large and growing number of studies of safety culture, relatively little has been published on how our understanding of cultural influences on safety can be used to improve or intervene in safety issues.

A few exceptions do exist. Roughton's and Mercurio's '*Developing an Effective Safety Culture: A Leadership Approach*' (2002) is one example, Florczak's '*Maximizing Profitability with Safety Culture Development*' (2002) is another. Both these books emphasize the role of leaders in 'shaping' cultures and the way training and the principles of Behavior-Based Safety (BBS) can serve as tools for improving safety cultures. Both also seem to be based on an 'iceberg model' similar to Heinrich's (1931) famous model.[1] In this, safety is equated with the absence of 'unsafe acts', involving a marked emphasis on employee behaviour and human error. Both these books are rooted firmly within an engineering approach to safety culture. In a more academic vein, Cooper's '*Improving Safety Culture: A Practical Guide*' aims to develop a more holistic model of safety culture. However, like Roughton and Mercurio and Florczak, Cooper, too, stays within the traditional approaches to safety improvement through auditing, training and behaviour based safety.

None of the three publications I mention take into account the anthropological origins of the concept of culture. This means they are somewhat blind to a number of limitations that apply to the 'improvement' of cultures, limitations that are inherent in a more social constructivist view of culture. In this chapter, I shall discuss these theoretical limitations, ethical and institutional constraints that may affect attempts to change cultures, as well as the possibilities that do exist for improving safety by means of a cultural approach.

---

1   Heinrich's model introduces a ratio between minor and major accidents, where minor accidents/incidents and unsafe acts are seen as underlying precursors of major accidents. This introduces the logic that safety management should aim to eliminate all minor accidents/incidents and unsafe acts, which, in turn, will be sufficient to avoid major accidents (see Heinrich 1931). This way of viewing the relationship between minor events and major accidents has been heavily criticized, most recently in the investigation report following the Texas City accident (Baker 2007).

Is it Possible to Influence Safety Cultures?

While the issue of cultural change has been largely ignored in research on safety, it has been extensively debated by *organizational* researchers. There is a wide spectrum of positions among organizational culture scholars regarding the possibilities of cultural change (see Ogbonna and Harris 1998 for a useful review).

The popular management literature remains highly optimistic on this issue, (e.g. Peters and Waterman 1983), but it tends to view cultural change in a rather mechanical way: Managers make some intervention effort and the members of the organization respond to this stimulus (Alvesson 2002). There is little discussion of how employees interpret or relate to intervention efforts.

Academic approaches tend to be far more modest with respect to the possibility of cultural change (e.g. Alvesson 2002; Ogbonna and Wilkinson 2003; Alvesson and Sveningsson 2008). The reasons for these differences are to be found in the way different authors conceptualize culture. In other words, the possibilities of and limitations to changing culture are to a large extent based in theoretical considerations. To limit the discussion, I will not discuss in detail the various positions within the field of organizational culture research. Rather, I shall focus on the limitations to cultural change that follow from the theoretical conceptualization of culture I have advocated in this book.

In Chapter 3, I argued for a social constructivist perspective, viewing culture as the product of the day-to-day interaction between the members of a given community. This perspective introduces at least two limitations on the way cultures may be influenced.

Firstly, large organizations will be culturally differentiated. They will include many cultural units; several groups of people interacting. This means that attempts to influence culture in large organizations will almost by definition need to involve *multiple* and *different* activities for effecting change .

Secondly, since cultures are produced locally through day-to-day interaction, culture is also *reproduced* and *changed* through social interaction. This means that 'one-size-fits-all' approaches, such as organization-wide 'culture campaigns', are very unlikely to have lasting effects and any attempts to influence culture that do not relate change efforts to the local interaction and everyday reality of employees are very likely to fail.

Because it is produced, reproduced and changed through everyday interaction, *not* by strategic decision-making, culture is not something managers can easily modify. It will, however, be possible to change the 'growing conditions' of culture, which in turn may lead to cultural change, but the results of such changes will always be unpredictable.

Should Cultures be Changed? Some Ethical and Institutional Considerations

In addition to the theoretical limitations, there are some important ethical and institutional aspects which serve, or rather *ought to* serve, as constraints for managers or researchers aiming to influence culture in an organization. These constraints are related to issues of power and were introduced in Chapter 4.

The ethical considerations concern the limits to an organization's extent of control over its employees and the fine line between organizational development and manipulation. The limits of any organization's control over its employees cannot be described as a fixed boundary. This is particularly true when it comes to safety, where there is a tendency for organizations to aim at a greater level of control over employee behaviour. Schlumberger, for example, has started to view attitudes to safety as a 24-hour business. Through their global driving programme, they provide mandatory driving lessons to reduce the risk of their employees being injured in traffic.[2] Implicit in such an approach is the ambition to train and socialize employees into having certain attitudes that will affect both their job *and* their personal life. Many of us might say that it is a good thing that a company cares about its employees and makes efforts to keep them alive and well. Personally, I have no doubt that such change efforts are made with the best of intentions. However, if one takes a more power-oriented perspective such approaches may look rather different.

In Chapter 4, I related Lukes's (1974) three-dimensional model of power to the study of safety culture. His third dimension of power, the power to influence the construction of meaning, is of particular relevance when it comes to cultural change. Translated to the realm of safety, attempts to influence culture will most often involve changing people's perception of what is dangerous and what is safe. This may include a model of what causes accidents with the aim of increasing employees' tendency to put safety first and propagating certain values with regard to how work processes should be performed. The logic behind such approaches seems to be to 'improve' employees' behaviour by trying to inculcate management values towards safety into the operational parts of the organization. This involves building consensus around a particular set of values regarding safety.

The methods involved in most of the large scale campaigns[3] toward safety culture improvement rely heavily on the principles of Behaviour Based Safety (BBS), as described by Krause et al. (1990).[4] This coupling is somewhat surprising, considering that a cultural approach and BBS have very different theoretical foundations and are considered by some to be mutually exclusive perspectives (see Tharaldsen and Haukelid 2007). This combination of culture perspectives and BBS is also where efforts to 'improve' culture begin to stray over the border of organizational development toward manipulation. The fusion

---

2   See Schlumberger's website for more information about the Global Driving Programme (http://www.slb.com).

3   Statoil's Safe Behaviour Programme (the Colleague Programme), Norsk Hydro's BuddyCheck and Shell's Hearts and Minds Programme can be seen as examples (Ryggvik 2008).

4   The relationship between safety culture and BBS have previously been discussed by DeJoy (2005) and Tharaldsen and Haukelid (2007).

between cultural perspectives on safety and the BBS doctrine of improvement involves a top-down, managerialist approach to safety improvement that suggests the dilemma pinpointed by Sejersted (1993), aiming to brainwash employees, *and at the same time* treat them as individuals. This is particularly the case where building consensus around the 'right' values is presented as a common good without opening the issue up for debate. The goal is to create a certain set of values and attitudes without asking those intended to adopt them about *their view* on the issues. Such an approach will always run the risk of trying to influence employees almost subliminally, without them actually being aware that they are being influenced.

We should also ask another question in relation to safety. Why are the values and interpretations of managers with regard to safety necessarily the right ones? How do managers know what is best, situated as they mostly are at the 'blunt end' of the organization, usually far away from direct exposure to risks? There is a growing body of literature on organizational learning with regard to safety that emphasizes the importance of having *requisite variety* in perspectives on risk and safety (e.g. Rochlin 1993; Weick 2001; Weick and Sutcliffe 2007; Nævestad 2008). The notion of requisite variety refers to having *multiple* framings and interpretations of a situation or a problem. This constitutes a better basis for learning and imagination about what may go wrong than where there is only one, dominant, view. From this view, aiming to create one, all-embracing culture by 'streamlining' the values and perspectives of an organization may actually reduce its ability to learn.

Of course, I am not claiming that all management attempts to change the way work is performed are unethical and illegitimate. Safety requires a high degree of control over and coordination of the processes involving risk, which means that management and leadership are essential to upholding and improving levels of safety. Any management's view on safety is, however, a view from the top. It is not the only view, and not always the correct one on safety related problems. Attempts to enforce one meaning of safety at the expense of any others can therefore reduce the organization's ability to uphold and improve safety.

A top-down, managerialist, approach to safety culture could be particularly problematic in societies characterized by low power distance (Hofstede 1984) and strong labour unions, such as the Scandinavian countries. The principles of Behaviour Based Safety are firmly rooted in a North American management tradition and the institutional context of US working life where the BBS approach allows managers significant leverage to govern and lead as they see fit without necessarily involving labour unions or safety delegates. It becomes problematic when transferred to the Scandinavian countries and others with traditions of participative management and labour unions. The BBS approach to safety culture bypasses these institutional arrangements through glossing over the fact that what is in managers' interests is not always in those of the workers. Ignoring and undermining the role of labour unions could actually be directly harmful to safety. As Ryggvik (2003) has argued, the Norwegian labour unions have important functions in the preservation of safety because they serve as institutionalized whistle-blowers, especially in the petroleum industry. So we have a paradox; the quest for a unified safety culture results in bypassing one of the most important organizational safety valves.

This underlines my contention that the development of safety cultures should not be based on an assumption of organization-wide agreement and harmony. On the contrary, in searching for a safety culture we should be looking for multiple opinions. The existence of conflicting views on safety can actually serve as an example of the requisite variety considered to be essential to stimulate an organization's learning abilities (Westrum 1993). A culture which influences safety positively is thus *not* necessarily one which is homogenous and conflict-free but one in which there is enough headroom to deal with opposing views in a constructive manner. This requires a more democratic approach to safety development.

The Principles of Action Research – Co-generative Learning and Employee Participation

Structured upon the theoretical limitations and ethical constraints that apply to efforts to change cultures, we need an approach that can serve to facilitate understanding between different groups within an organization, facilitate the horizontal and vertical flow of information and bridge the gap between the formal and informal models of work. I argue in the remainder of this chapter that the principles of action research can be instrumental in achieving these three objectives in safety interventions.

The concept of action research is inextricably linked to the thinking of Kurt Lewin, who had a strong influence on industrial sociology in general and the tradition of industrial democracy in particular. Lewin conceptualized change as a three stage-process, consisting of an unfreezing or unlocking of existing structures, changing them and finally freezing new structures into a permanent state (Greenwood and Levin 1998). Rightly, in my view, this view of social change has been criticized for being too sequential, overlooking the continuous character of learning processes (ibid.). Within the field of safety, which is highly regulated and relies heavily on rule-based management, the notion of unfreezing or unlocking of structures is, however, of great importance. In the case studies presented in this chapter, I shall show that the ability to unlock some of the organizational 'rules of thumb' about how to deal with safety issues is key to successful organizational safety interventions.

As a tool for accomplishing democratic social change, the principles of action research (AR) involve three basic elements: investigation, action and participation. The process is described in the following way by Greenwood and Levin:

> AR is social research carried out by a team encompassing a professional action
> researcher and members of an organization or community seeking to improve

their situation. ... Together, the professional researcher and the stakeholders
define the problems to be examined, co-generate relevant knowledge about
them, learn and execute social research techniques, take actions, and interpret
the results of actions based on what they have learned. (Greenwood and Levin
1998: 4)

The investigation phase involves a broad mapping of the organization which may
be based on a variety of sources of information, depending on the situation in
question – quantitative data (e.g. accident and incident statistics, survey data),
interviews, focus groups, ethnographic fieldwork and document analysis, to name
some of the alternatives.

This initial problem definition usually forms the starting point for so-called
search conferences or dialogue conferences.[5] These are conferences or workshops
that ideally include actors representing all the relevant interests in a problem. A
search conference is a methodology where planning, creative problem solving and
concrete action are integrated activities (Greenwood and Levin). The conference
proceeds as a mix of group discussions and plenary sessions. Search conferences are
usually facilitated by third-party consultants or action researchers making sure that
the voices of all groups are heard. The goal is to continue the problem formulation
by bringing out as many as possible of the interpretations and 'worldviews'
involved and to discuss possible solutions to the problems identified.

Note that it is *not* an explicit aim of a search conference or dialogue conference
to *influence* cultures. There *is* an explicit focus on informal processes, as the
emphasis is on the various interpretations of a problem, but the goal is to facilitate
understanding and trust *across* cultural settings. This is partly why the learning
processes of action research are labeled 'co-generative'. The learning that takes
place is not achieved through teaching or influence, it is created through a collective
activity of dialogue. This is one of the key strengths of the action research approach
in efforts to improve safety. It also illustrates that the ideals of action research rest
on democratic principles and the participation of all stakeholders in the solving of
a problem.

This brief introduction to the main principles of action research does not
do justice to the thinking that underlies the approach and it is superficial as it
conceals the fact that there are several different schools within the community
of action researchers. In order to show how these principles may be useful in
safety interventions, we need some real-life examples. The case studies presented
in this chapter will give a more vivid account of how the general principles of
action research translate into practice. The reader is also referred to Greenwood

---

5   It is common to separate between search conferences and dialogue conferences,
as search conferences are more directed at problem definition than dialogue conferences
which are seen as tools for more continuous improvement processes (e.g. Levin and Klev
2002). I choose not to separate between these forms of conference as they are similarly
structured and organized.

and Levin (1998) and Reason and Bradbury (2008) for a thorough introduction to action research.

The Role of Researchers in Safety Interventions – Ethical Considerations

In the context of organizations, action research usually involves an emphasis on employee participation and balancing the power inequalities between managers and the managed. However, action researchers rarely reflect explicitly upon whether, as researchers, they themselves enter into power relations in the field being studied. This is a weakness since there are at least two forms of power relationships action researchers should address. Firstly, action research is usually funded by the organization in which the research process takes place. This means that representatives of the organization have control over some of the funding and field access on which the researchers are dependent in order to do their job. This is an obvious source for possible power alliances between action researchers and managers in the client company. An example might illustrate this. In oil-producing countries, a great deal of the research on safety issues is, in one way or another, linked to the oil and gas industry. This means that those who research safety in these countries are heavily dependent on having some kind of access to the industry. This dependency is exaggerated by the tendency for large oil companies to grow ever larger. A good example is Norway, probably one of the countries where safety researchers are the most dependent on the petroleum industry. Here the two major oil companies merged in 2007. This means that as a researcher you no longer have two possible 'customers' or partners but only one and in this way the market power of the large oil companies becomes dominant. Control over the scarce resources of research funding may thus involve researchers forming power alliances with company management 'against' the workers.

Another form of power involved in the relationship between researcher and members of the organization is that of being able to frame and form problems by connecting them to scientific knowledge. This can be a source of power by giving legitimacy to one party's interpretation of a problem. The claim to be an expert can also create a risk that the *researcher's* interpretations of a problem are favoured at the expense of those of the 'problem-owners'. Specialist knowledge may thus act like blinkers, hindering researchers in making important observations.

Interestingly, both of these ethical constraints have been stressed by Todnem (2009) in his description of experiences with action research seen from the viewpoint of a major oil company. The fact that these constraints are also recognized on 'the other side of the table', so to speak, serves to illustrate that they are not just hypothetical constructs. Rather, they form very real threats to the ideals of action research, which emphasises the empowering of the *weaker* parties in a problem situation.

The point of the discussion about the power relationships that may exist between researchers and the organizations we study is not to imply that action research is the same as business consultancy. My point is rather to show that aspects of researchers' professional integrity should be recognized as an important constraint in our efforts to change organizations, especially when it comes to culture. In fact, the most important practical function of safety researchers may reside in their ability to question the dominant beliefs and approaches of the practitioners.

I now turn to two case studies which will provide empirical examples of how the principles of action research can be applied in real life situations. The first case study presents the lessons from a successful safety intervention aiming to reduce the gap between procedures and practice on supply bases serving oil installations in the North Sea. The actual intervention was not made by action researchers, but the principles of action research nevertheless guided the intervention process. This example illustrates that the ideals of action research can inform all organizational change process, not just processes where external researchers take part. The second case study is a more conventional action research approach and tells the story of a long-term development programme aiming to improve safety in processes involving service vessels in the petroleum industry.

Case Study One - Reducing the Gap Between Procedures and Practice[6]

This case study addresses the problem of not following procedures by focusing on *compliance* rather than violation. The aim is to shed light on the conditions that can facilitate a balance between models of work as laid down in formal procedures and the way it is actually carried out in normal operations. We shall do this by evaluating a study of a change process that aimed at improving the use of procedures on Norwegian offshore supply bases. Evaluating the effectiveness of safety interventions is an important task of safety research since it can yield information about the usefulness of different intervention strategies (Shannon et al. 1999). Unless we evaluate the effectiveness of interventions we will overlook one of the key tasks of safety research, to contribute to improving safety in the settings studied. The study presented here aims to extract some success criteria from a major revision of the procedures governing work on the offshore supply bases and to show how some principles of action research can contribute to the implementation of safety interventions.

Background

Work procedures are a part of all organizational design. In organizations with operations subject to high risk, operating procedures play an important role in the systems designed to control risk. Lind (1979) defines procedures in the following way:

---

6   This section is partly based on an article co-authored by Petter Almklov and Jørn Fenstad (Antonsen et al. 2008).

> In general, a procedure is a set of rules (an algorithm) which is used to control operator activity in a certain task. Thus, an operating procedure describes how actions on the plant ... should be made if a certain system goal is to be accomplished. (Lind 1979, cited in Dien 1998: 181)

According to this definition, procedures are protective mechanisms against human error. In addition to this control aspect, other functions could also be included such as coordination between tasks and the accumulation and diffusion of organizational experience. Procedures also have a legal function in fixing responsibility. Employees have a responsibility to perform tasks in a way that does not jeopardize safety. If procedures are deliberately violated, this may imply negligence that, in turn, may involve legal liability.

It follows from this that safety procedures are the very backbone of safety management systems. In fact, all safety management systems are based on the assumption that people will follow the procedures most of the time (Hudson et al. undated.). If this assumption is wrong there is little value in having procedural systems at all, so therefore there should be no doubt that employees' use of and adherence to procedures is of critical importance to the study of organizational safety.

Even so there is surprisingly little research on the topic. Some researchers have analysed the reasons why employees violate procedures (e.g. Reason 1990; Mason 1997; Lawton 1998; Hopkins 2000; Hudson et al. undated; Karwal et al. 2000), but there are few attempts to propose solutions to this problem (e.g Hale 1990; Bax et al. 1998; Dien 1998; Laurence 2005). The research on the topic is also quite unsystematic, which is probably why Hale and colleagues (2003) summarize the research on safety procedures in the following way:

> The cry is for better rule use and management, but there are still not very clear ideas of how and of how to avoid the many pitfalls in producing workable safety rules. (Hale et al. 2003:2)

That there are discrepancies between the work practice prescribed through safety procedures and the way work is actually carried out should not come as a surprise to anyone. Sociological studies of work very often reveal that workers tend to create their own informal work procedures and that these can be very different from the formal ones (e.g. Gouldner 1954; Crozier 1964; Suchman 1987).

Formal procedures and informal work systems   The reasons why procedures are violated are many and complex (Reason 1990). However, both researchers and safety managers have a tendency to focus on the individual aspects of violations, such as employees' attitudes, commitment and competence. Helmreich's (2000: 746) view on error as resulting in the 'physiological and psychological limitations of humans' is one example of this. Although he recognizes the role of contextual factors such as workload and 'flawed decision making', the emphasis remains on the individual workers and their limitations. Such views rest on the assumption that total safety will be achieved if all employees comply with the rules at all times. This is a simplification of a multifaceted problem symptomatic of bureaucratic, top-down approaches to safety management. One of the most far-reaching consequences of such rationalistic approaches is that planning of operational practices is separated from the people performing the work (Morgan 1986). Procedures are developed solely by experts who are not involved at the operational level and have few options for participation or feedback on how useful they are. Very often this leads to a gap between 'work as imagined' and 'work as actually done' (Dekker 2006), as procedures produced from afar are less able to cope with challenges and unforeseen events.

This is where informal work systems start. Once established, they involve an erosion of control which, in turn, can lead the organization into a practical drift (Snook 2000) towards the boundaries between safe and unsafe practice. Moreover, the development of informal work systems often involves a cultural differentiation within the organization in question. Lysgaard (2001) for instance, observed that the industrial organizations he studied in the 1960s were characterized by a marked division between the operational workers and the administrative staff. The shop floor employees drew a cultural line between 'us' and 'them', the office employees. According to Lysgaard, the employees' collective functioned as a buffer to the demands of a rigid bureaucratic system. In another classical study in the sociology of the subordinate, Crozier (1964) shows how employees' collectives may be a medium for resistance to organizations' control systems. Both these studies, although conducted half a century ago, contain important information for those studying the relationship between procedures and practice. The friction between, on the one hand the workers' collective and the technical/economical system on the other, is an important factor in understanding the degree of commitment to an organization's safety management system and also how safety interventions can be successfully implemented.

Procedures and organizational learning   The body of procedures regulating an area of work sometimes seem like a jungle. In complex workplaces procedures may be in conflict with others both within the company's own system and also with those of contractors. The sheer number of procedures may thus constitute a barrier towards their use. A safety manager we had previously interviewed illustrated this when he complained that 'we have the best procedures in the world – the problem is that no one uses them!'

There are several mechanisms that contribute to a steady growth in the number of procedures. One is the complexity of work. The attempts to make the procedures as realistic as possible may serve as a major reason for an ever-increasing number of them. There are institutional mechanisms, too, that contribute to the generation of new procedures. In the aftermath of accidents and incidents, regulatory authorities usually demand some kind of action to be taken to prevent recurrence. A common and highly visible way of satisfying such demands is to create new procedures.

This is tantamount to saying that safety management systems have a tendency to become so-called *loosely coupled systems* (March et al. 1976; Meyer and Rowan 1977). In a loosely coupled system, the formal organizational structures, such as rules and procedures, do not arise from the demands of the concrete work activities. Instead, they reflect the myths of their institutional environments (Meyer and Rowan 1977: 341). This means that new procedures are generated on the basis of influence from forces outside the organization such as regulatory authorities and management literature and not as responses to the practical needs of the organization itself.

Procedures are often amended to prohibit actions that contributed to creating hazardous situations. As a consequence they have a tendency to become increasingly restrictive (Reason et al. 1995, in Mason 1997: 292). This illustrates that the development of procedures represents a form of organizational learning where the aim is to translate information about past failures into future improvement.

Within the theory on organizational learning, it is common to distinguish between 'single-loop' and 'double-loop' learning (Argyris and Schön 1996).

Single-loop learning is a form of instrumental learning in which an organization seeks to improve already existing strategies and processes It is a form of learning which 'connects detected error – that is, an outcome of action mismatched to expectations and, therefore, surprising – to organizational strategies of actions and their underlying assumptions' (Argyris and Schön 1996: 21). The basic assumptions and underlying logic of the activities involved is not questioned. The term 'single-loop' thus refers to the feedback loop between observed effects of an action and the strategies behind it. Improvement of procedures most often bears the marks of single-loop learning as it involves the refinement of already known strategies and courses of action.

Single-loop learning is commonly contrasted with double-loop learning. This is a more 'advanced' form of learning. In addition to instrumental improvement, double loop also calls into question the underlying logic of the system. The term 'double-loop' thus refers to the feedback loops between the observed effects of an action, the strategies which created it, and the underlying logic of the system. Related to safety management, double-loop learning would involve questioning whether safety could be managed by principles fundamentally different from the current ones, e.g. how procedures are to be used and whether formal procedures are the best way to manage safety.

Bourrier's (1998, 2005) studies of compliance with procedures in French and American nuclear power plants in many ways highlights the importance of rethinking the forms of learning involved in safety management systems. She found that there were three crucial ingredients for a successful match between procedures and practice. First, there should be feedback from the lower to the upper tiers of the organization. Second, the adjustment of procedures should be based on the views of those directly involved, particularly front-line employees, and third, the time interval between worker feedback and implementing changes should be as short as possible. This way of adapting procedures to practice, instead of trying to change practices to fit procedures, challenges the underlying assumptions of traditional safety management in which there is a fundamental division between planning and performing work.

Bourrier's studies point to a number of mechanisms of learning which display some, but not all, characteristics of double-loop learning and that it is possible to reduce the gap between procedures and practice. As I shall show, her findings are of great relevance to this study.

Case Description

The primary task of onshore supply bases is to provide the logistics services necessary for the shipment of goods and equipment to offshore oil and gas installations. These are owned and staffed by businesses offering their services to several oil companies operating on the Norwegian continental shelf. The oil company in this case operates several supply bases along the coast of Norway. The supply service business carries out all physical handling of goods between receipt and the wharf side, i.e. registration, handling and packing.

Work processes on the supply bases include several hazardous operations such as the loading and unloading of supply vessels involving lifting with cranes heavy containers, drill pipes and other equipment. There have been serious incidents with crane lifting operations on the supply bases, including a 17-ton chain cable falling from the crane onto the wharf. Such incidents involve severe material damage and have the potential to kill or seriously injure personnel.

The work carried out on the supply bases also impacts on safety in other parts of the logistics chain. In particular, supply operations have experienced perpetual problems with objects e.g. mislaid tools, stones and flakes of rust falling from containers when these are lifted on or off the supply vessels. This becomes a major problem when containers are lifted from supply vessels onto the oil installations and the crane is up to 50 metres above the deck of the supply vessel. Any objects falling from this height have the potential to kill the seamen working on the vessel deck. There have also been incidents where because of faulty packing heavy equipment has fallen out of containers when they are opened, again with severe risks for those doing the opening. As well as these work-related incidents, the oil company in question experienced a serious situation when one of their installations was minutes away from a gas blowout that could have had very serious consequences.

Examination of all these incidents revealed deviations from procedures, some very serious, and this focused a great deal of attention on the company's use of procedures. The regulatory authorities were, of course, concerned and demanded that the company initiate some sort of measure to improve compliance with internal procedures. This marked attention to procedures provided the backdrop for the changes that were made in the safety management systems within the supply bases.

Objectives of the change project   The purpose of the change project studied here was to improve the use of and compliance with procedures at the supply bases. This general objective was broken down into several operational goals:

- To have only one document regulating the supply base operations.
- The document should define the relationship between both customers and principals.
- The document should contain the existing requirements for the performance of all tasks involved in supply base operations.
- The language should be easy to understand.
- The operative workforce should be involved in the development of procedures.

The procedures, previously dispersed among 15 different documents, were now collected into a single one and overlapping procedures were merged so that the number was reduced. In addition the quality of the document was improved by keeping the language simple and precise.

The outcome of the changes was a document consisting of about 25 pages. The procedures formulated in it, from now on referred to as WR1, are meant to cover all work processes on all the company's supply bases.

The research project was tasked with evaluating the potential effects of these changes. The logic behind the changes was quite simple. If there is to be any chance of procedural compliance, the users must know where to find the procedures which must contain relevant information that the users are able to understand. By the changes, therefore, the formal work requirements would be expressed in the real-life context of work and worker involvement was seen as a prerequisite to achieving this goal. This was the ideal scenario but the operative workers were not greatly involved in the initial formulation of the procedures. Once a draft of the document had been formulated, however, all employees were invited to comment on and suggest improvements to it. So while the ideal of democratic participation was thus *not* followed in the initial phases it was compensated for by a very direct form of participation at the implementation phase. This direct, as opposed to representative, participation is seen as the key success factor in the change process.

Methods and Design

The case study is best described as evaluation research. The scientific merit of evaluation research has traditionally been controversial since it is, in some ways, 'commissioned work'. The problem to be addressed is defined by someone other than the researchers and the organization funding the research may have political or economic interests in the results of the study. Thus, the scope and direction of evaluation studies are influenced by forces other than 'pure' research interests. While this is certainly something researchers need to be aware of it is probably the situation in most organization research. The organizations involved in research will always be stakeholders with an interest in its outcome. In this particular case, the company's motives for our involvement as researchers in the evaluation were that they wanted it carried out by someone independent of the organization. Since the respondents and interviewees were people working for a subcontracting company, it was considered important to reduce the risk that the evaluation might be perceived as some sort of safety audit on behalf of the oil company. Hence it was the clear obligation of the research team to provide external, independent input to the evaluation. While the representatives of the oil company were certainly a part of the evaluation, there were no attempts to influence its *results*.

It is also worth emphasizing that while narrowing the traditional distance between researcher and the research subjects represents a challenge, it also offers significant advantages. In particular, evaluation studies often enable good access to the organization involved and first-hand knowledge is, in turn, crucial to finding the right informants, asking the right questions and making sound inferences.

Design of the study   Many would perhaps argue that, in terms of internal validity, studies involving the evaluation of the effects of some change should be designed as experimental or at least with some sort of before-and-after design. This has not been done in this study. Admittedly, this puts some restrictions on the inferences that can be made from it as it is not possible to control the variation of factors other than the changes in procedures. However, there is quite a considerable amount of existing knowledge on the use of procedures in the Norwegian petroleum industry. For instance, biannual questionnaire studies of the trends in risk level on the Norwegian continental shelf reveal several problems related to the usability of procedures (cf. Petroleum Safety Authority 2005). The same deficiencies, as well as serious violations, were also found in the investigation into the gas blowout on the Snorre A platform (Schiefloe et al. 2005). This illustrates that there have been persistent problems related to the usability of and compliance with procedures in the Norwegian petroleum industry. There is little reason to assume that the supply bases differ from the rest of the industry in this respect, since they are subject to the same procedural regime. The weaknesses in research design were compensated for by combining quantitative analysis with extensive qualitative data. A quantitative approach was thought necessary in order to reach all the users of the procedures in question. A qualitative approach was necessary in order to shed light on what causes compliance or violation. These are complex social phenomena which are difficult, if not impossible, to analyse through survey methods alone.

Ideally, the perceived effects of the changes should be accompanied by a reduction in the accident statistics. This was not the case in this study. If anything, there seems to have been an increase in the number of incidents in the year after the changes (Antonsen 2006). The purpose of this study, however, is not to determine the relationship between procedures and safety statistics. The goal is to shed light on the mechanisms that facilitate compliance with procedures. Whether complying with procedures is actually the safest way to perform work is not the main issue here. The emphasis is on how employees perceive and relate to procedures, not how procedures eventually influence safety statistics.

Data material   As already indicated, the data material includes both quantitative, questionnaire, and qualitative, interviews, data. The quantitative data target was 98 employees at the supply bases. The survey was in the form of a self-completion questionnaire, consisting of around 35 questions, sent to all employees whose work responsibilities were regulated by the procedures. The response rate was 80 per cent.

The qualitative data material consists of individual and group interviews with a total of 18 informants. The interviews were semi-structured and each lasted approximately one hour. The interviews were recorded and later partially or fully transcribed.

Results

The results of the evaluation are described. After a brief presentation of the overall perceived effects of the changes in WR1, I turn to the respondents' views on key characteristics of the procedures as well as the success factors behind the process of implementing WR1.

Figure 7.1 shows the respondents' evaluation of the overall effects of the changes in WR1. Around 70 per cent of them perceived that the changes had improved safety. Although 12 per cent of them said the changes introduced through the new WR had no effect on safety, the overall response from the employees is highly positive.

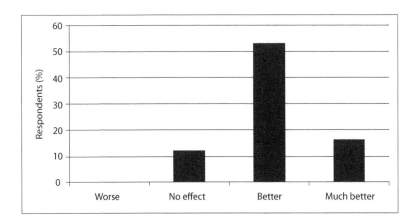

Figure 7.1   **'All in all, how have the changes in WR1 affected safety?' N=80**

The interview data reveals that the main factor behind this effect was perceived to be the simplification of the body of procedures into one document:

> If you can make things simpler, it will be an improvement. And my view is that they have simplified the procedures into one uniform 'bible', so to speak.

One document which was widely known and recognized among employees in different divisions, occupational groups and organizational levels provided a single point of reference for all activities on the supply bases. However, there is very little point in merely gathering the various bits of documentation in one place if nobody understands or adheres to it. Some additional steps need to be taken in order to reduce the gap between what is stated in procedures and done in practice. These steps regard the characteristics of the procedures as well as the characteristics of the process of implementing them.

Characteristics of procedures: Comprehensibility, accessibility and accuracy   If procedures are to have any chance of being relevant to real work situations, there are some prerequisites to be satisfied. First, the employees must understand the language used, a question of comprehensibility. If one does not understand the language used, one cannot understand the content of the procedure. Second, the information must be accessible to the employees, a question of accessibility. There is no point in having good procedures if no one knows where to find them. This was a real problem before the simplification. Third, the procedures must contain a good description of the relevant tasks, a question of accuracy. A procedure which is based on or contains faulty or inaccurate information cannot, and rightly so, be expected to be adhered to in real work situations. As shown in Figure 7.2, the respondents largely view WR1 as satisfactory with regard to these three characteristics.

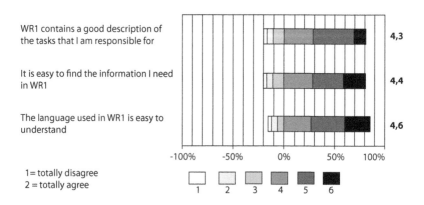

Figure 7.2   Respondents' evaluation of the content of WR1. Per centages and mean values. N=87

Comprehensibility, accessibility and accuracy are not the only conditions for compliance with procedures but they are, nevertheless, necessary ones as they reduce the probability of unintentional violation. As pointed out by Laurence (2005), voluminous safety management plans and procedures characterized by complex and complicated language cannot be expected to 'connect' with operational workers. Procedures should therefore be developed with the operational workforce, not experts and regulatory authorities, as the main target group.

Perceived effects of the changes   Reducing the complexity of the procedures was seen to make them more accessible to the employees. Over 60 per cent of the respondents acknowledged that they either agreed or totally agreed that the changes in WR1 had made relevant information more available to them (Figure 7.3).

Gathering the procedures into a single document reduces the employees' 'costs' of finding the relevant procedures, e.g. navigating, sorting out irrelevant procedures:

> There is quite a bit of knowledge in WR1, related to the work we do. ... It is nice to have the document to look up and read a little bit about it. As opposed to before, when it was terribly difficult to find one's way [in the procedures]

Having one document as a 'bible' for all operations also has the unintended effect of facilitating the printing of paper copies of it from computer-based systems. Strictly speaking, making paper copies is a procedural violation as all updates of procedures should be published through the computer-based system. However, making paper copies is, in many ways, something that has to be done so as to have the procedures available, especially for those employees with no immediate access to computers. Having a single document containing all relevant procedures can thus be a way of making it easier for operational employees and those with limited computer skills to gain access to procedures.

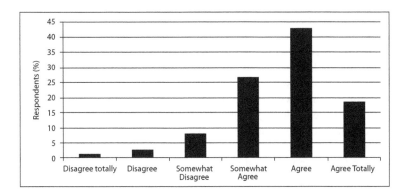

Figure 7.3   **'The changes in WR1 has made relevant information more available'. N=75**

In addition to, and probably also a consequence of, the improved access and availability of relevant procedures, the employees saw the simplification of them as a clarification of what is demanded in work performance (Figure 7.4).

Most saw this as one of the main positive effects of the changes; only 13 per cent disagreed with the statement.

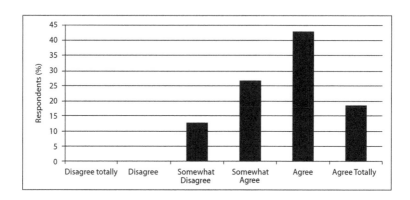

Figure 7.4   'The changes in WR1 have made it easier to know the demands that pertain to work performance'. N=78

This clearer picture of what is mandatory in performing work is probably related to an increase in knowledge of existing procedures. Although no objective assessments of the extent of knowledge about procedures has been made, it seems highly likely that better access and availability would contribute to its increase. The interviews also supported the impression of a high level of knowledge of the procedures in WR1. As well as better access and availability, it is also very probable that the great focus on the new procedures at the time of the investigation increased knowledge about them.

Having one single document also provides a description of the work processes that functions as a common point of reference between different divisions, groups and organizational levels and serves to clarify the lines of responsibility between the different work processes.

The ultimate objective of simplifying procedures is to improve employees' compliance with them. Figure 7.5 shows that the majority of employees report that compliance with procedures has improved after the changes in WR1.

This shows that reducing complexity can be an important strategy to close the gap between formal and informal work practice.

This does not imply that procedures are no longer violated. In fact, 44 per cent think that the procedures in WR1 are sometimes violated. Although most of these violations are attributed to customers or suppliers, also subject to regulations in WR1 such as requirements for the packaging and securing of goods, violations also occur on the premises of the supply bases. When asked which procedures are violated most, informants point to those that are seen as imposing excessively detailed restrictions on the workers. To comply with these procedures requires a great deal of time and effort and the workers do not share the judgement of risk that underpins the procedure.

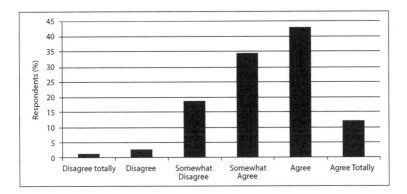

Figure 7.5 **'Compliance to procedures has increased after the changes in WR1'. N=74**

To the employees on the supply bases, the new document implied they had more protection against pressure from the oil and gas installations, undoubtedly the more powerful parties in the logistics chain. As Figure 7.6 shows, the majority of respondents reported that the new procedures had made it easier for them to handle pressure from customers and suppliers.

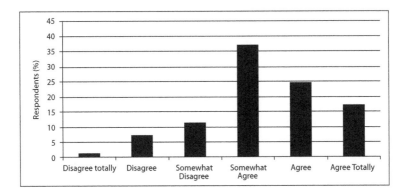

Figure 7.6 **The changes in WR1 have made it easier to handle pressure from customers and suppliers. N=74**

In the interviews, this aspect of the new procedures came out as involving a sort of empowerment compared with the previous situation in which the rules governing activities were less clear. This serves to illustrate that rules and regulations need not always be seen as control mechanisms. In fact, having a common standard for work performance seems to have reduced some of the strain and friction between different parts of the logistics chain:

> It's a great tool. Almost everything is described in there. It is nice to be able to say, 'these are the rules here, and here is a copy of them', instead of having to explain and argument and getting angry customers. ... 'We use it against the oil installations. Against them that try to take short-cuts, and try to make us take short cuts'

These benefits from the new procedures seem especially important for the larger supply bases. The more complex the activity, the more the workers rely on formal procedures to handle pressure from suppliers and customers.

Some dissenting views   The positive picture calls for some qualifications. There were certainly some dissenting views, evident in the number of respondents who chose to disagree with the statements in the questionnaire. The interview data provides some information about the sources of this dissatisfaction. Three stood out as particularly important. First, some of the employees performing specialized tasks were dissatisfied with the new procedures because information about their tasks had been removed. This had the practical effect of making less guidance available for these tasks and, perhaps a symbolic effect, implicitly reducing them to a sort of peripheral activity. Second, some of the interviewees were fundamentally opposed to procedures as such. These were probably also the interviewees with the least knowledge about the new procedures. Third, one of the supply bases had a long history of conflict between employees of the oil company and of the supply service company. The climate for cooperation between these two groups was certainly not ideal and seems to have had a negative influence on the implementation of the new procedures.

Implementing the Procedures: Some Success Factors

Although important, reducing the number and complexity of procedures is only the first and in many ways the easy step in the process of improvement. Getting them accepted in the operating parts of the organization can be challenging, especially in those where informal work procedures have become established.

The implementation of the new procedures was carried out in the following way. A draft of the document was presented to all employees working on the site, whether those of the contractor of the oil company. The draft was discussed, and the employees had the opportunity to comment on ways to improve it. It was then modified and officially introduced as part of the current regulations of the organization. Importantly, the implementation process did not end at this point. There was a continuous focus on the new procedures in weekly HSE meetings and other arenas regarding health, safety and environment. In addition, all employees were required to sign a statement that they had read WR1. Although some saw this as an attempt to transfer the juridical responsibility of accidents to the workers, the majority felt that the request to sign highlighted the employees' responsibility to work in accordance with procedures. A third strategy in the implementation was the development of matrixes, showing which procedures were the most important for each function on the supply base. This is an effective way of helping the employees to differentiate between the procedures they are expected to have thorough knowledge of and those it is sufficient to be acquainted with.

Taken together, these strategies were successful in diffusing knowledge about the new procedures. One informant summed up the process in the following way:

> There has been a lot of activity related to the implementation of the new procedures. A lot of people have been involved. For once, they [the oil company] have been very thorough. At least as I have seen it on my base, everyone that should know of the document has been invited to review the document. The implementation has been very good compared to what we have been used to before.

> [Interviewer:] Previously they would have just handed over the finished document? – No, previously we would discover by coincidence that there had been introduced some new document regulating our work!

The employees responded very positively to the implementation process. The opportunity to discuss the form and content of procedures before they were formally implemented, as opposed to being presented with decisions already made, was especially perceived as an important success factor.

Simplicity and involvement   What are the lessons to be learned from the processes studied? As already indicated, the simplicity of the document and the worker involvement in the implementation process are seen as the two main success factors. I now turn to the question of *why* these factors are crucial for reducing the gap between work as an assumption and work as actually done.

That simple and accessible procedures have better chances of being used than those that are complex and inaccessible is self-evident. For the supply bases, having a single document that describes the relevant work processes with no irrelevant contingencies reduces the costs in time and effort involved in using the procedures. There is a risk, to be avoided, that the process of simplification might make procedures too general to offer any real guidance on work performance.

Reducing the complexity of procedures is the easiest part of the endeavour to improve the match between work as designed and work as performed. The real challenge lies in their implementation. Most safety researchers and practitioners have heard workers complain about procedures that are useless because they are developed by 'pen pushers' with little or no knowledge of the practical work. In some instances, workers see the introduction of formal procedures as an attempt to devalue their practical knowledge (Knudsen 2005). Such reactions make it clear that procedures are often seen as something imposed from higher levels and I suspect that this is often correct in the way procedures are being implemented in organizations. This is one aspect of what previously was labelled a loosely coupled system (Meyer and Rowan 1977), characterized by a fundamental distance between the *creators* of procedures and those who *use* them and is one of the main factors contributing to the divide between work as designed and work as performed.

The success of the change process studied here can be attributed to the attempts to tighten the coupling between procedures and practice. This was done through involving employees at different levels which gave the process a strong local anchoring. The role of worker involvement in the implementation process can be seen in relation to Bourrier's (1998, 2005) findings, which were described above. All the factors critical for compliance highlighted by Bourrier, feedback, worker involvement and closeness in time between feedback and change, were present in the process studied. And yet the managers of the oil company did not achieve this result through the pursuance of a conscious and preconceived strategy. In fact, in the initial phases of the study the oil company representatives felt the level of involvement to be so low that there was no need to include issues of worker involvement in the investigation, as they already knew that there had been virtually none in the traditional sense. They only agreed reluctantly to include questions of worker involvement in the survey and interviews, so it came as something of a surprise when the workers expressed such positive views on the level of involvement. Although no operational workers participated in the project group that produced the *draft* of the document, the workers felt they had had influence on the *final* document by being invited to comment and discuss the draft.

Although there is a long tradition in Norwegian working life of worker participation, this particular case involved a broader form of involvement than is common in the Norwegian oil industry. The usual form of worker involvement in the industry is to have safety delegates and union members as representatives on committees and in project groups whose tasks are likely to affect the workforce. Clearly this only involves a few selected members of the total workforce and might not be viewed as broad involvement of the end users.

The implementation of WR1 was characterized by a more direct form of involvement than just of representatives; it sought to involve *all* end users of the document. One of the informants emphasizes the part played by dialogue in the process:

> Compared to other oil companies, I think this one has a more humble attitude towards their subcontractors. ... They don't just say: 'This is how it's going to be – fix it by tomorrow!' We can always discuss things, there is plenty of headroom.

So even if the customer is always right, this customer is willing to listen to other points of view.

Previous studies have shown that such two-way communication is important for the ability of an organization to learn and thus for safety (Antonsen et al. 2007). Commitment-based approaches to safety work, e.g. worker involvement, have also been shown to correlate with employees' commitment to safety and trust in management (Barling and Hutchinson 2000). From an operational worker's perspective, a procedure can be seen more as one possible resource among others for solving the given task rather than a blueprint of how to carry out the work. Because there is a tendency to stereotype and simplify other people's tasks when they are viewed from a distance, how the practitioners see the job is probably the best way to judge the level of detail needed (Suchman 1995).

The risk of oversimplification As we have already noted, some interviewees felt that simplifying the documents had led to the removal of important information thus bringing about unexpected and unwanted consequences. This raises an important question about the desirable level of detail in work procedures. Should procedures be detailed prescriptions for task performance, or is it sufficient just to make the broader functional requirements clear? Analysis of our research shows that it is impossible to give one general answer to this question. Whether simplification can be seen as an improvement depends on the nature of the tasks involved. For some, e.g. those that are performed rarely, are highly complex and/or require a high degree of coordination, it is probably necessary to have detailed descriptions and regulations. More routine tasks do not need the same degree of direction and can be expressed in more general functional requirements. These two approaches to controlling and coordinating work activities are the extremities of a continuum that ranges from imposing control and standardization of the detail at one end and allowing autonomy and professional judgement at the other. Determining the level of detail control along this axis ought to be based on consideration of the nature of the tasks involved. This further underlines the importance of obtaining worker involvement, since the operational workers are, in many ways, the experts on the practical tasks and the context in which they are performed.

Attitudes and procedures Focusing on the properties of procedures and the way these are implemented does not mean we can just disregard individual attitudes towards procedures. There were certainly individuals who were opposed to procedures in general and who expressed frustration that procedures did not allow them to use common sense and perform work in what *they* saw as the best way. Resistance to procedures is not synonymous with a disregard for safety. What these employees protested against was the formalization of safety. They saw safety as resulting from sound professional judgement and practical work experience, not as something that can be assured through written procedures. This means that the continuum between detail control/standardization and autonomy/professional judgement can also serve to describe individual attitudes towards procedures. Since individual attitudes and perhaps also group cultures are likely to vary along this axis, it will be impossible to reach a situation in which there is a complete consensus on the role and quality of work procedures.

Conflict and change   The third of the dissenting views presented related to the level of conflict in the organizations studied and its influence on the change process, which seems to have been significant. The supply base with the highest level of conflict did not achieve as positive results as the other bases. The conflict had originated in connection with some organizational changes over a decade ago when some employee groups were transferred from the oil company to the supply service company. This involved the loss of some fringe benefits and was seen as a downgrading in terms of status, causing great dissention among them. Some of these traces of conflict still persist and seem to pose difficulties in building trust and cooperation between some of the groups involved in the activities of the supply base. Clearly, a poor climate for cooperation between management and operational levels does not facilitate management-induced change processes.

This illustrates that organizational change processes rarely follow a predictable, strictly rational pattern. Historical issues like old conflicts can inject considerable noise into the implementation process which, in turn, may influence the way change processes are translated into real work practices.

Why is worker involvement important in implementing procedures?   At this point in the argument it should be clear that worker involvement is regarded as the key condition for achieving a better match between procedures and practice. The question that remains is *why* is it so important?

Worker involvement provides four main benefits that are crucial in building acceptance and adherence to safety procedures:

1. When workers are consulted in the development of procedures it is likely to increase their ownership of them. This sense of ownership, in turn, is important to the legitimacy of procedures since it reduces the chance that they are perceived as instruments of external control, developed by people with no practical work experience. A higher degree of dialogue in the development of procedures can certainly contribute to bridging some of the cultural differentiation that often exists between safety managers and shop-floor workers.
2. Worker involvement in developing procedures provides safety managers with an opportunity to communicate the *intentions* underlying the procedures. This actually forces safety managers to consider their own intentions, i.e. how the procedures should be interpreted, how workers should use them and how the formal procedures relate to real work performance. This is a form of learning for safety managers since it challenges the traditional division between the planning and performing of work. By engaging in a

dialogue about the contents of procedures and how they should be used, safety managers will be more directly confronted with the relationship between procedures and practice and so obliged to consider the workers' point of view on procedures.

3. Establishing a dialogue between the makers and the users of procedures can also be instrumental in diffusing knowledge about their contents, role and logic. Since knowing what procedures are about is fundamental to compliance with them, this factor should not be underestimated.

4. Beyond their mere prescriptive function, procedures are part of an inventory of tools or resources which the workers use and this transcends mere compliance and violation. For example, it seems that WR1's status as a single paramount reference point made it a better tool for handling relations with customers and suppliers. Also, it was easy to print a copy of it, even though this was frowned upon by management, and so handier to carry around; none of this is irrelevant to compliance. Tools must be created with a view of the context in which they are meant to be used and of the hands in which they are supposed to fit. Since procedures also function as a tool for management, e.g. fixing responsibility and demonstrating commitment to work safely, there is always the danger that this may lead to an undesirable and uncontrolled growth of procedures that makes them less usable for the operators. Worker involvement at the point where procedures are constructed is a good way to gain input about how they may work in practice and to create tools that actually meet the challenges posed by the situations in which they are used.

These four beneficial effects of worker involvement may seem trivial and self-evident. After all, the importance of worker involvement has been stressed in theories of action research and organization development for decades. In the field of safety management, however, these insights seem to have been largely neglected. Bureaucratic control systems keep on being developed without questions being asked as to whether they actually work as intended. Therefore, there seems to be a need to inculcate the ideals of worker involvement within the field of safety management.

Are the improvements permanent?   Despite the problems I have described, the results of this study show that the simplification of the requirements governing work and the involvement-based implementation process has led to a distinct improvement in the workers use of and adherence to procedures. But is this a lasting effect or will things return 'back to normal'?

To a large extent this is a question about whether the company is able to maintain the simplicity of the document and the sense of ownership created during the implementation of WR1. There are some aspects of safety management systems that pose challenges for both simplicity and involvement. As I have already indicated, learning from incidents, a major goal of all safety management systems, often involves regulating ever more aspects of work through procedures.

There are also some tricky aspects of procedures and work process descriptions that sometimes inhibit discussions about them. For us, this was best illustrated by discussions that took place at an early stage of the project. To get input from the workers and to ensure their involvement in the evaluation of the changes, representatives from both workers and safety managers were invited to a workshop. Something that kept coming up in this workshop was that the worker representatives would claim that not all problems could be solved by procedures. But when the workers were prompted to give examples, the procedure team was always able to 'solve' the problem by referring to the existing system. The example the workers had raised could either be included in working practice by the feedback mechanisms in the governing documentation or it was classified as a breach. The key point that there were things that did not quite fit into the system was obscured because in every example of this it could be shown that it did fit.

The relationship that the documentation has to its surroundings is such that every piece of the *praxis* in the workplace can be fitted into it, although, always piecemeal. Hence, the argument that the complexity of the work in the real world is too great to fit into a document is difficult to exemplify. Why can the document not cover the whole work when it evidently *can* cover every piece of it?

Unless these tendencies towards producing more and more procedures to cover every aspect of work are addressed, the simplification and usability will be only temporary. One of our informants, a safety manager, also underlined this point:

> When you have such revisions of the body of procedures, you often end up with removing the procedures seeking to standardize solutions to eventualities for which there really was no need for procedures in the first place.

This underlines that the match between work as designed and work as actually performed should be monitored continuously, rather than only through periodical revisions.

Addressing the Gap Between Procedures and Practice: Opportunities for Building Resilience

Why is it important to address the gap between procedures and practice? Investigations about the match between procedures and practice represent a form of 'organizational inquiry'[7] (Argyris and Schön 1996: 11) which, in turn, can form the basis of organizational learning. Moreover, the lessons learned by the company display some of the traits of double-loop learning. By looking

---

7   'Organizational inquiry' is defined as a form of doubt and reflection arising as a result of a 'problematic situation,' triggered by a mismatch between the expected results of action and the results actually achieved' (Argyris and Schön 1996: 11)

beyond the relationship between individuals and procedures and focusing more on system properties and the context of work, the inquiries into the relationship between procedures and practice involved questioning the way they conceived of procedures. Instead of only asking 'how do we ensure that employees comply with procedures', safety managers asked the question 'how can we make procedures in a way that facilitate their being used?' Here, the subject matter of reflection is not only how to design a good work process but also on *how* work process designs are best used in governing the work processes themselves. In other words, addressing the relationship between formal and informal aspects of work may involve posing questions about the underlying principles for organizing.

One should be careful, though, not to overemphasize such traits of double-loop learning in the case studied here. While the safety managers involved seem to have posed some fundamental questions regarding the way procedures should be developed and implemented, the contents of the procedures and their status in governing work processes remains the same. In this area, improvements are still incremental which illustrates that the relationship between single-loop and double-loop learning is not dichotomous. Learning processes may display characteristics of both forms. In the case studied here a form of double-loop learning seems to have occurred with regard to the *processes* of development and implementation of procedures. At the same time, the content and status of the *product* is changed according to the principles of single-loop learning.

The ability to reassess current processes, strategies and models of risk is a central part of the concept of resilience (Hollnagel et al. 2006). Addressing the gap between procedures and practice can be seen as an opportunity to create resilience. It is the only way of making 'silent deviations' audible (Tinmannsvik 2008). Some of these deviations from procedures can be smarter and safer ways of performing work while others may involve short cuts that compromise safety. In any case, the process of examining these silent deviations is an essential step in asking a larger question, how well are the organization's existing safety management strategies matched to the demands of the organization's work activities and organizational environment?

Asking this question raises another. Is it possible to *eliminate* the gap between procedures and practice? The answer is that this would be impossible since the number of local and situational variations that constitute the context of work is indefinite and therefore impossible to describe within formal procedures. Some sort of translation and adaptation between plans and situated action therefore seems unavoidable (Suchman 1987). For exactly the same reason elimination of the gap between procedures and practice is quite undesirable; for organizations to perform safely they must be able to adapt to local variations. We can see this particularly in crisis situations where the ability to respond vigorously is of critical importance. When a crisis emerges there is rarely time to consult procedures and formal documents. The ability to make exceptional violations may thus be safety critical. In addition, crisis situations may come as fundamental surprises for which there simply are no procedures or emergency plans.

Coping with such situations hinges on the ability to improvise and act creatively (Sætre 2006). The very autonomy and improvisation that organizations try to reduce in order to prevent accidents from occurring may, paradoxically, be their greatest asset in order to tackle the unforeseen events that do happen. This hints at a possible tension between strategies of *anticipation* and strategies of *resilience*.

The fact that there could be conflict between prescribing for safety in day-to-day operations and ensuring it in worst case scenarios has important implications for the management of safety; one should be very careful not to rely too heavily on detailed control and strict procedures as this could adversely affect safety when unforeseen events happen(Weick et al. 1999).

In this I agree with Rasmussen (1997). Instead of organizations concentrating on controlling behaviour by attempting to stop deviance from pre-planned procedures, they should rather make explicit and known the boundaries of acceptable behaviour . There is an important balance to be achieved in safety management between strategies of anticipation and strategies of resilience to achieve both reliability in normal operations *and* resilience towards unforeseen events. This is an issue that is implicit in recent writings on resilience engineering (Hollnagel et al. 2006), but which nevertheless has not been dealt with openly in previous research and theory and should be a key question for future safety research.

Concluding Remarks on the Relationship Between Procedures and Practice

The case study highlights some fundamental conditions that need to be satisfied if procedures are to have any significant influence on actual work practice. The changes to the procedures made them more concrete and more accessible to the workers. This, in turn, resulted in greater compliance and more active use of them. The success factors behind these improvements are to be found both in the simplicity of the documentation and, more importantly, in the broad and direct form of worker participation that characterized the process of implementing the procedures. The main conclusion to be drawn from this analysis is that addressing the gap between 'work as imagined' and 'work as actually done' can create opportunities for organizational learning as it allows safety management systems to be viewed from the sharp end. This form of probing into the relationship between work, hazards and management systems is an important prerequisite for building organizational resilience.

Case Study Two – Improving Safety Through the Principles of Action Research[8]

Organizational strategies that seek to foster employee commitment are generally thought to have positive effects on key organizational qualities such as learning ability, effectiveness and productivity (e.g. Weisbord 2006; Parker 2006). The goal of the rest of this chapter is to show that same logic can be applied to organizational safety.

An action research group associated with the Norwegian University of Science and Technology collaborated closely with a Norwegian oil company from 2000 onwards. The objective was to improve the safety of the supply services that provide the transport and emergency preparedness necessary to operate the offshore installations. The long-term nature of the collaboration, as well as the closeness of the relationship involved, provided very good access to the organization for the researcher group. In addition to conducting three surveys, about 40 interviews and analysing the company's accident statistics, the research group participated in numerous meetings at different levels in the company. The arguments of this section draw upon all these sources of data.

The section consists of three main parts. First, I provide a rough description of different approaches to safety management. Second, I give a brief account of what can be characterized as a lock-in situation in the organization studied. As I describe in further detail, a lock-in situation is one where an organization is 'stuck' in certain ways of thinking and dealing with safety challenges. In the third part I describe and discuss the measures initiated by the action research group with the aim of 'unlocking' the organization.

Different Approaches to Managing Organizational Safety

The report *Organising for Safety* by the Advisory Committee on the Safety of Nuclear Installations (ACSNI 1993) is the first comprehensive review of the literature on organizational safety (Hale and Hovden 1998). This report divides the history of safety management into three phases:

In the first phase the approach was largely *punishment-based*, characterized by what can be described as a 'scapegoat mentality'. When incidents occur, the main focus is to find those responsible and punish them. In this approach accidents tend to be seen as the results of unsafe individual acts and little attention is paid to whether the unsafe act has anything to do with systemic conditions. The approach has its roots in classical behaviourist psychology.

The second phase was characterized by what is labelled 'prescribing in advance' (ACSNI 1993:3), and may be called a *bureaucratic approach*. Detailed rules and work procedures are the main regulating devices in this approach, a firmly

---

8   This section is partly based on an article co-authored by Lone S. Ramstad and Trond Kongsvik (Antonsen et al. 2007).

structural one. Safety is seen as a management responsibility and as something that can and should be directed and controlled by management. The principles underlying this approach can be traced back to Heinrich (1931), but general management theories from the 1960s and 1970s also exercise a strong influence. The bureaucratic approach represents a development on the first phase since it recognises the impact of organizational practices on safety. It relies on achieving control through measures and policies calling for compliance (DeJoy 2005).

The third phase is the *cultural approach* developed out of an increasing interest in the informal aspects of organization and which is reflected in the approach presented in this book. As I have already argued, this approach is highly compatible with practice-based methods such as action research. Nevertheless, surprisingly few attempts have been made to study organizational safety through action research. Richter (2003) and Richter and Koch (2004) are honourable exceptions, as they adopt principles of action research in their studies of safety culture. Although these studies are useful, they fail to address a question of vital importance; *In what ways are the principles of action research useful for improving organizational safety?* This is the primary research question to be addressed in this second case study.

I described the above-mentioned approaches to safety as phases but I ought to make it clear that within organizations, to varying degrees, one can find all three of them even though the bureaucratic approach is still the managerial method most frequently applied in order to improve or maintain the level of safety in an organization (Barling and Hutchinson 2000). As I will argue, the bureaucratic approach comprised a significant element in the lock-in situation in the organization that provides the empirical basis for this article.

The Story of Safety Development in a Distributed Offshore Logistics Activity

This case is based on empirical work conducted by several researchers over a period of four to five years. The case story documents a project aimed at improving safety among the service vessels in the maritime operations of a Norwegian oil company.

The offshore oil industry in the North Sea is mainly serviced by three types of vessel. Supply vessels are used to transport goods to the oil and gas installations, specially equipped vessels operate anchor-handling activities and stand-by vessels take care of the emergency preparedness.

The supply services conduct the lion's share of the activity, which can be seen as a value chain (Porter 1985) or a logistics chain. This chain describes the physical and sequential flow of products or goods through different parts of the organization, or across different organizations. The logistics chain of the supply vessels is complex, and as previously mentioned, several organizations and actors are involved, the ship-owners' offices, the ships' crews, the oil company with its different departments and the crew on the installation. Although, or perhaps because, there are a number of organizations working together the activity in the logistics chain is to a great extent controlled and coordinated by the oil company. In addition to communicating directly with the oil company in their daily operations, the vessels are subject to an extensive set of rules and regulations. Also, the nature of the contract between the ship-owners and the oil company grants the latter considerable influence over the ship-owners' safety management system.

The complexity of the logistics chain poses major challenges for safety management. The next section describes the way safety was initially dealt with.

The lock-in situation   From the very start of the oil industry in Norway in the 1970s, work on the service vessels has been associated with considerable risk. From the mid 1990s, however, the oil company's figures for lost time injuries (LTIs) showed a significant increase. Between 1996 and 2001 the number of LTIs nearly tripled. In addition, the number of collisions between service vessels and the oil company's installations doubled every year. From one collision in 1997, the number rose to three in 1998, six in 1999 and twelve in 2000. These collisions represented an increased risk of injury to personnel as well as severe damage to property and of potential shut-downs in the production of oil and gas on the installations. The actual and potential financial losses associated with collisions between vessels and installations were considerable.

Response to these developments was firmly rooted within the bureaucratic approach. Investigations carried out after injuries and serious incidents often revealed failures to comply with the existing rules and procedures. Instead of seeking to shed light on the underlying reasons as to *why* the rules were not complied with, the common response was to introduce yet more rules and procedures. Thus new or modified work procedures became a largely 'standard' management response to injuries and incidents. This response can be described as rational in that the efficiency of the organization is based to a large extent on the principle that work can be standardized and pre-programmed. Standard operating procedures are then adjusted to cover each specific situation as it arises. Such a prescriptive safety control mode is shown in Figure 7.7.

Reason (1997) argues that the body of safety procedures increases through additions which reduce the scope of action required to perform tasks effectively. Problems occur when the scope of permitted actions shrinks to such an extent that violation of procedures becomes routine practice. Another problem is that procedures to ensure safe work operations suffer from a lack of requisite variety. In virtually all hazardous operations the variety of possible unsafe behaviours is much greater than that of required safe behaviours. The requisite variety of the procedures necessary to govern safe behaviour will always be less than the possible variety of unsafe situations.

Yet another problem is that the procedures are developed by experts who have no involvement at the operational level. Morgan (1986) argues that one of the most far-reaching consequences of the rational approach is that the planning and pre-programming of work are separated from the people performing it.

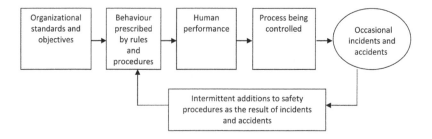

Figure 7.7   A feed forward process control, based on procedures with intermittent additions

Adapted from Reason (1997).

This separation between planners and operators was the case with the service vessels where the procedures were developed without any real participation from the workers. The relationship between safety managers and workers was thus to a great extent one of linear, top-down communication. This is characteristic of the bureaucratic approach to safety which tends to treat safety as extraneous to work practice (Gherardi and Nicolini 2000).

This one-directional situation can be described by a very simple model, commonly used to describe a linear communication process (Figure 7.8).

Note the model does not include any form of feedback, so the sender knows very little about the recipient's interpretation and use of the message.

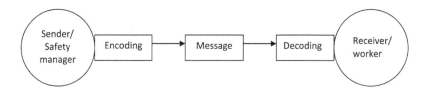

Figure 7.8   Linear model of communication

Adapted from Adler et al. (1983).

The organizations involved found themselves in a situation with a complex and growing body of work procedures. At the same time, the number of collisions and injuries continued to increase. It became obvious to the managers of the logistics chain that the traditional response of achieving safety through increased standardization was no longer effective, causing them both concern and frustration.

With hindsight, the safety managers related their problems to the one-directional mode of communication inherent in their approach to safety management. One of the managers summed up the situation like this:

> We had no arenas where we could meet the captains and address their views, where they were alone and not dependent upon others. We had a previous arena called the 'safety forum' which included captains, platform managers, personnel, safety representatives, lots of people. But the only thing that came out of it was bickering, and it was the same issues being discussed over and over again. ... we needed to change tactics to foster enthusiasm, and not be pointing fingers like we used to.

Deteriorating accident and incident rates combined with a feeling of having neither the perspective nor the methods to turn the situation around, constitute the essence of what can be defined as a lock-in situation. Recognizing this, the managers of the maritime operations department brought in external expertise in the form of our research group, some members of which had previously been involved in other research projects in the organization; probably an important factor in the oil company's choice of collaborator.

The next section describes the action research activities which contributed to new collective practices and improved safety in the organization.

Unlocking the Organization

The action research approach   In order to get an overview of possible causes and measures, in 2000 the oil company carried out a survey in cooperation with our research group. A sample of 434 individuals was drawn from the crews of the service vessels, 76 per cent of whom completed the survey.

The main conclusion was that an efficiency improvement project in the logistics chain starting in 1997 had led to increased physical strain on the crews and caused attention problems. At the same time the crews were experiencing pressure from other parts of the logistics chain. In particular, the supply vessel crews reported that they felt pressured to go through with loading/unloading operations alongside the platforms irrespective of the weather conditions. As indicated in Chapter 6, the vessel crews commonly saw themselves as the 'underdogs' of the logistics chain who could be pushed around to satisfy the whims of the platform crews.

In 2001 the oil company asked our research group to carry out a development programme where the ultimate goal was to improve the safety performance of their maritime operations involving the service vessels. Our research group had previous experience with practise-based methods in organizational development. An important assumption in the development programme was that safety could be improved in a more fundamental and lasting manner by involving the community of practice whose safety was at stake. Stakeholder involvement was thought to be able to improve motivation, encourage development of more effective solutions and measures, and also foster a more holistic view among the stakeholders of the safety challenges they were experiencing. The main stakeholders in this case were the crews of the service vessels, but other personnel groups in the logistics chain became more important as the programme developed.

Action research (Greenwood and Levin 1998) was considered an approach that met the basic premise of achieving stakeholder involvement. The ships' captains were considered key personnel and important stakeholders in safety questions since they carried the overall responsibility for the activity on board. We therefore established an arena – 'Captain's Forum' – where the action research process took place. We arranged regular search conferences where safety questions were addressed. The first search conferences were devoted to reflections upon why accidents happened and what could be done to reach the ideal of zero accidents. Later search conferences focused on how different actors in the logistics chain could cooperate in order to improve the safety results. Hence, personnel from the onshore supply services and the offshore oil and gas installations also became important participants in the search conferences.

Our research group carried out different research activities as input to the search conferences. We analysed historical data and earlier accidents involving the service vessels and created a database. We also carried out new surveys and interviews involving the crews. The results from these activities were presented to and jointly interpreted by the stakeholders and the research group. These discussions resulted in a list of specific preventive actions. In later search conferences these actions were then evaluated jointly. This circle of establishing and evaluating measures is still running.

In addition to search conferences, 15 other measures aimed at improving the situation were put into effect by the oil company. These included individual, physical, organizational and communication measures (Hansson et al. 2004),[9] several of which originated in the discussions in the 'Captain's Forum'. The search conferences thus served as arenas both for assessing and defining problem areas related to safety and for brainstorming and discussing solutions to these problems. As shown in the next section, the action research approach appears to have been effective.

Effects of the measures   The safety level of the service vessels has increased significantly since the interventions. From 2001 the annual collision frequency has been between 0 and 1, which is a dramatic improvement on the 12 collisions in 2000 (Figure 7.9).

The frequency of personnel injuries per million working hours decreased from 13.8 in 2001 to 2.6 in 2005 (Figure 7.10).

---

9   Among the measures were increasing the number of navigators on supply vessels, technical measures to improve noise reduction and navigator simulator training (Hansson et al. 2004).

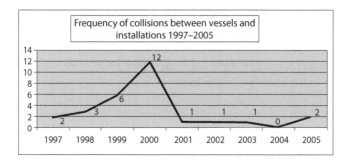

Figure 7.9 **Development in the frequency of collisions between service vessels and installations in the period from 1997 to 2005**

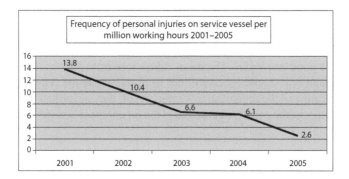

Figure 7.10 **Development in the frequency of personnel injuries on service vessels per million working hours in the period from 2001 to 2005**

It is impossible to establish with certainty that these improvements are due to the measures taken in the safety development programme. However, an evaluation of the programme concluded that there are few alternative explanations (Hansson et al. 2004). Although the oil company has contracted a small number of newly built ships, the technological standard of the remaining fleet is still largely the same. Technological reasons are thus unlikely to explain more than a limited proportion of the overall improvement, which is too radical to be attributed to coincidence.

The crews on the vessels seemed to agree that things had changed considerably in terms of the way safety was managed in the logistics chain:

> I have to hand it to them [the oil company], they have changed in recent years. It seems obvious that someone has pushed them a bit too, over the last few years. They have gained a better understanding of the work that we do.

It is worth noting that the informant traces the changes back to an increased managerial understanding of the work at the shop floor level, which is one of the major purposes of action research. Another informant expresses a similar argument, emphasizing that there has been a shift away from what bears strong resemblance to a 'punishment' approach:

> ... I think that it [communication] has changed a bit from what was before. If an accident or incident does happen, they sometimes jump at us. I think that has been taken more calmly, it is no longer that harsh reaction, but at the same time they take things seriously and try to do something about it.

This can be seen as a shift away from a blame-oriented towards a more learning-oriented approach to accidents and incidents.

The crew members' rating of the oil company also showed significant improvement after the safety development programme was introduced. Respondents to the survey assessments in 2000 and 2002 were asked to state their level of agreement with the following statement: 'Compared to others, this oil company is a good employer'. The mean values rose from 3.89 in 2000 to 4.46 in 2002. Although one should be careful not to overestimate the relevance of such general indicators to the organization's safety level, they nevertheless serve as an important index of the perceived changes in the oil company's safety management approach.

The role of action research The lock-in situation was characterised by a considerable distance, both physical and psychological, between those who designed and maintained the safety management system and those who were supposed to work in accordance with the regulations built into this system. A consequence of the bureaucratic approach to safety management is that the safety management system takes on a life of its own. The focus on designing and maintaining the system becomes so all-consuming that the question of whether it is working as intended is completely ignored. This failure is no doubt associated with the general tendency among various regulatory authorities to perform system revisions that aim to assess which safety systems exist rather than whether these systems are actually working as intended.

The action research helped to break up this pattern in a number of ways:

1. Increasing worker participation and involvement instrumental in enhancing employees' trust in management, and their commitment to safety (Barling and Hutchinson 2000). Search conference methodology strongly emphasizes worker involvement. The action research approach increases the workers' ownership over the definition of and possible solutions to safety challenges. This is fundamentally different from the control-oriented approach as it accepts that the 'official' way of doing things is not necessarily the safest way to perform risky activities.

2. Ensuring a more unified approach to safety in the logistics chain. The bureaucratic approach focuses mainly on procedures and organizational structures. It pays little attention to organizational culture. In this case, the vessel crews perceived themselves as 'underdogs' who were under pressure to compromise safety in order to comply with the platform crews' requests. This hints at the existence of subcultures in the logistics chain the main goal of which is to ensure that the oil and gas production proceeds as smoothly as possible and tends to single out the platforms as the most important part of it. Such an emphasis has cultural implications, creating differences in status between platform and vessel employees which then contribute to the perceived efficiency pressure. What is important here is that these kinds of culturally induced risks are extremely difficult, if not impossible, to assess through traditional safety management methods such as formal incident reporting. The assessment of such risks demands an approach which involves all relevant stakeholders and which is thus both unified and practice based.

3. Promoting two-way forms of communication. As one-way communication was a major element in the lock-in situation, two-way communication is a major part of the unlocking-process. This form of communication introduces the concept of feedback into the linear model (Figure 7.11).

The search conferences were designed as forums for collective learning, collective reflection on practice and collective inquiry and they provided crucial arenas for such two-way communication. As already indicated, the search conferences led to a variety of new safety measures. Since to a large extent these measures are based on practical experiences from operational work, they are likely to be better adapted to practical work tasks and are also perceived as more legitimate by the workers.

The fact that the workers' feedback is quickly translated into action serves both a manifest and a symbolic purpose. Manifest in that it can provide safety improvement/learning *without* having to wait for incidents or accidents to happen. Symbolic in that it sends an important message to the participants that safety is important and that everyone has a role to play in safety improvement. Such symbolic functions should not be underestimated within the field of safety improvement.

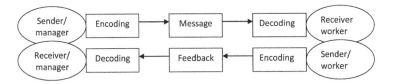

Figure 7.11   Interactive model of communication

Adapted from Adler et al. (1983).

To criticize some aspects of the traditional strategies of safety management does not imply that work procedures are irrelevant to safety. The pivotal question, however, is whose expertise underpins these procedures? Procedures without a firm basis in local safety knowledge are likely to be seen as constraints rather than opportunities. My criticism is of the top-down oriented bureaucratic approach to the production of procedures rather than of procedures as such.

The oil company also attributes improved efficiency, as well as the clear improvements in safety levels, to the safety development programme. Reportedly, the collaboration on safety improvement has given the different units in the logistics chain a better understanding of each other's work responsibilities and challenges. This in turn has been said to improve the overall cooperation and communication and thereby the efficiency in the logistics chain. The connection is difficult to prove empirically, but it may also be argued that a cultural approach to safety improvement is likely to have effects beyond improved safety levels because it targets communication and mutual understanding as central aspects of the improvement processes. These aspects of organizational behaviour affect the performance of complex human structures in more ways than just safety, suggesting that the traditional view that the relationship between production and safety must inevitably be adversarial may not be quite so self-evident.

The action research approach has been instrumental in providing the means for unlocking the organization. However, an important question remains: is it now fully unlocked?

Is the organization unlocked?   In establishing the search conferences as regular arenas for safety communication and learning, action research can stimulate the organization towards a more interactive form of communication. However, there are few 'quick fixes' when it comes to organizational development and it is also the case in this organization.

In 2004 we conducted a survey and 20 interviews to assess safety culture in the logistics chain. The results showed that some of the problems, especially with regard to work procedures, were still present. The oil company's way of managing safety was still the object of a certain amount of confusion and frustration. For example, many still called for increased worker participation in the management of safety. Nearly 17 per cent of the respondents in the survey disagreed with the statement that they had sufficient influence on the development of procedures and regulations.

This was also reflected in the qualitative material. The following quotation from a first officer is a good illustration that the oil company still has work to do in terms of incorporating local knowledge into its safety management approach:

> I am highly critical to the stressing of procedures and checklists. ... But that doesn't mean that I don't take safety issues seriously. ... Focusing on safety, putting it on the agenda, and focusing on the use of protective equipment, among other things, that is very important. But going from that to controlling every

detail, telling workers how they should behave and work in every situation, that's too excessive in my opinion.

Although some of the problems relating to a bureaucratic safety management approach are still present in the organization, to a greater extent the management now knows about and appreciates the workers' frustration. Organizational change is a complex process and one cannot expect immediate transformation. The managers of the logistics chain do show a growing willingness to abandon the control-oriented perspective. For instance, the oil company has agreed to let the shipping companies take over much of the investigation after incidents and accidents.

This brief account of the current situation indicates that action research can provide the necessary, but not the sufficient conditions for unlocking an organization. Action research provides the tools for local development processes, but it takes continuous and goal-oriented efforts for these processes really to 'take hold' in the organization.

Concluding Remarks on the Role of Action Research

In this chapter, I have tried to show how an action research approach can contribute to bridging the gaps between different cultural units within an organization, to facilitate the vertical and horizontal flow of information and to align formal and informal notions of work. With its emphasis upon worker participation, dialogue and learning, action research can represent a powerful approach to the unlocking of organizations. However, old habits die hard and the case studies presented here suggest that there are no quick fixes for improving the organizational aspects of safety. Nevertheless, by facilitating feedback communication and dialogue, action research can help create the conditions necessary for unlocking the organization. It is very clear that the principles of action research are not only relevant for researchers but for anyone who wishes to improve the organizational aspects of safety.

Recommendations for Safety Interventions

Since I have expressed a great deal of scepticism regarding the possibilities of cultural change, I have painted myself into a corner on how cultures may be changed. This means that anyone expecting a novel formula for influencing culture is about to be thoroughly disappointed. I have no prescription to offer. In my view, the conclusion of what I have presented so far is that culture is not something which necessarily can or should be influenced directly. However, we need to *understand* culture in order to achieve change. This is a very different way of viewing the relationship between culture and change than many existing approaches, particularly those influenced by the doctrines of Behaviour Based

Safety. It involves turning from a normative to a more descriptive stance regarding culture. Nevertheless, as cultures are always in the making, I believe that cultural change may be achieved as a 'by-product' of organizational change. But since culture is a slow-moving phenomenon, this will take time and require very persistent change efforts.

Therefore, while I cannot offer any recipe on how to change culture, what I can offer is some guidance about how change processes should be undertaken in order to be consistent with the approach advocated in this book. My recommendations in this area are influenced by the writings of Alvesson and Sveningsson (2008) and Amundsen and Kongsvik (2008).

First of all, organizational change involves a form of 'self-transformation' (Alvesson and Sveningsson 2008:176). No change has been achieved until there is a change in the practices of the organization. This means that the entire organization should be included in the change process.

Second, the goals of change processes should be moderate and relate to everyday realities. This follows from the constructivist perspective on culture. The very origins and nature of culture demand that change efforts must in some way connect to social interaction to have any hopes of success.

Third, there are no quick fixes. Organizational, not to mention cultural, change should be considered as a long-term project. All too often, change becomes ritualized in the sense that it is more motivated by a desire for change *itself* than a response to actual problems (Amundsen and Kongsvik 2008).

Fourth, combine aspects of 'push' and 'pull' in change efforts. Managers must of course take initiative and contribute to drive processes forward. However, real involvement requires that employees below management level are continually involved in dialogue about the objectives, means and progress of the process.

Fifth, be sensitive to organizational symbolism. To achieve lasting change it is far from sufficient just to express and explain the objectives of the change process. For instance, anyone can state that safety is the number one priority and that there is always time to work safely. Such slogans are void of any meaning until they are followed up through action and real-life priorities. Large gaps between the front of stage visions of change and the backstage actions and priorities will be very likely to derail change processes.

Sixth, be sensitive to what makes sense locally. Change efforts must connect to people's real life experience and day-to-day interaction. This also involves trying to understand how efforts to change are being perceived by different groupings in the organization.

Seventh, and most important, the goal of change should not be to create organization-wide consensus. It should rather be to create a common language and an understanding of the social realities of different groups; very different from aiming to create one, unified culture. On the contrary, the objective of change is more about looking at and understanding the differences between groups and, in turn, bridging the gaps *between* them, than reducing them to one single culture. This is particularly important when it comes to safety, as having multiple cultures may serve as a form of requisite variety, forming a better basis for learning and envisaging what may go wrong than where there is only one, dominant view. For example, the labour unions in the Norwegian petroleum industry see it as part of their mandate to challenge the companies' views about what is safe and what is not. With this you have a constant dialogue and questioning about the 'established truths' of management. If you streamline and tightly manage the organization's views and interpretations about safety, you will lose this safety valve.

Eighth and last, consider the need for change and how realistic it is that the goals and visions will be realized. As Amundsen and Kongsvik have stressed, managers often have unrealistic expectations about the outcomes of efforts to change and the time span needed to achieve them. A consequence of initiating multiple changes without considering the need for them and the realism of their objectives could be an organization that becomes resistant to change and cynical about management intervention efforts (Amundsen and Kongsvik 2008).

Chapter 8

Conclusions

The overall purpose of this book was to shed light on some fundamental theoretical, methodological and empirical questions regarding the relationship between organizational culture and safety. The general problem it addresses is: *How can a cultural approach contribute to the assessment, description and improvement of safety conditions in organizations?* The term 'cultural approach' applies to studying the way the cultural processes and traits of organizations influence the practice and levels of safety within them. The various parts of the book focus on different aspects of the general question I have just posed and in this final chapter I shall try to show how this book provides answers to it.

The main answer to the question of how a cultural approach may be useful to the study and improvement of safety in organizations is that it is necessary to shed light on the informal aspects of work and organizing. The study of organizational, including occupational, culture can tell us something about why some interpretations, actions or practices have meaning for the actors involved while others do not. When we take a cultural approach we can make visible the things otherwise taken for granted in the way we perform our tasks. Such an approach can provide information regarding the frames of reference through which information is interpreted and social reality is constructed, as well as the conventions for behaviour, interaction and communication.

A cultural approach means probing into some very fundamental properties and processes of an organization. This information, in turn, is essential for understanding the factors that contribute to the creation and prevention of both organizational and occupational accidents. It pertains to those 'irrational', or perhaps non-rational, aspects of social life that are hard to assess through traditional safety management or psychological perspectives. Taking a safety management perspective, the focus tends to be on the formal aspects of safety, while a psychological perspective emphasizes the role of individual perceptions, attitudes and values regarding safety.[1] A cultural approach, on the other hand, emphasizes the *informal* and *social* processes that influence safety. This is not to say that the cultural approach is in any way superior to other approaches to the study of safety. Cultural traits and processes certainly permeate aspects of technology, formal organization and individual perception, but one should be careful not to reduce all aspects of safety and organization to matters of culture. Issues of risk and safety are far too complex

---

1   It should be noted that much psychological research and human factors research also includes elements that are not related to individual perceptions, attitudes and values. The main emphasis is nevertheless on individual and cognitive aspects of safety.

to be explained by reference to one single factor. Among others, aspects of formal organization and technology are major determinants for the level of safety in an organization[2]. A cultural approach is therefore just one piece of the puzzle depicting the different mechanisms that influence the performance of a system, although it is an important one. As has been stressed by Hollnagel and Woods (2006), the ability to maintain control over the variability within the performance of the system is a major contributor to safety.

Control over the variability of performance requires a high level of knowledge about the way the system actually performs. This, in turn, requires knowledge about the informal conventions that influence practice, the assessment of which requires a cultural approach.

Some Recurrent Themes

Although the book addresses theory, methods and practice related to safety culture, there are some themes that have recurred throughout the book.

The Relationship Between Formal and Informal Ideals for Work Performance

The most conspicuous of these recurring themes is perhaps that all the empirical studies presented in the book deal in one way or another with the relationship between formal work requirements and informal notions about work performance. This relationship is probably one of the pivotal issues of social scientific safety research in general, and of safety culture research in particular. Hale (2000), for instance, sees it as the fundamental 'mission' of safety culture research to tell us why the structures of safety management are, or are not, working. When the empirical analyses presented here all point to a considerable degree of friction between formal work requirements and informal notions of work, this indicates an inherent dilemma of safety management. There is an apparent contradiction between formal procedures and local adaptation and improvisation, between the need for predictable operations on the one hand and the ability to respond to unforeseen events on the other. While these two strategies might appear contradictory they should not be conceived of as a dichotomy but rather as a continuum between extremities that need to be balanced in some way.

The continuum between rule-based management and local adaptation to some extent corresponds to the dilemma between centralization and decentralization identified by Perrow (1984). In short, Perrow's argument is that organizations whose activities generate a number of unforeseen events and non-routine tasks need to be decentralized in order to adapt to local challenges. At the same time, the ability to maintain control over a system requires centralization. While

---

2   See Schiefloe et al. (2005) for an outline of the various aspects of organizational aspects that may influence safety (the Pentagon model).

Perrow's dilemma is concerned with the properties of high-technology systems, the dilemma between centralization and decentralization is also of relevance for the settings studied here. One interpretation of the findings of this book is that the organization studied relies heavily on centralized safety management, while, at the same time, many of its activities are dynamic and non-routine. This interpretation is supported by the fact that many of the effects of the safety interventions in the cases studied in Chapter 7 are the consequence of a rethinking of the relationship between centralization and decentralization. In both the service vessel and the supply base cases, improvements in the levels of safety were achieved by giving up some centralized control and increasing the level of employee participation. This represents a form of decentralization which aims to attune the formal structures to the way work is actually carried out.

A rethinking of the way one goes about managing safety has been identified in this book as a key issue in improving overall levels of safety, indicating that effective safety management in an organization is not just a question of having adequate control mechanisms governing the work processes. On the contrary, the improvements described in the case studies seem to occur in instances where the boundaries between planning and doing, that are such a conspicuous feature of safety management systems, have been transcended. This points to a possible weakness in the way safety is managed in the organizational contexts studied here; to an undue extent it is equated to the detailed control of work processes. We saw this problem in the investigations of the Snorre Alpha incident and again as some of the background to the changes at the supply bases discussed in Chapter 7. In Westrum's (1993) taxonomy of the ability of organizations to learn to manage safety this way of doing it seems to be based on bureaucratic principles of governance. For instance, the flow of information regarding the management of safety seems to originate primarily from the top and is based on written communication through formal information systems. Of course, there are exceptions to this. The department responsible for the logistics activity involving the service vessels achieved significant improvements by taking measures which aimed to discontinue this bureaucratic mode of organizing. In all the empirical studies involving Statoil, however, we saw a propensity towards viewing safety management as predominately a question of control over behaviour. Organizational efforts in the realm of safety seemed to be more directed at prescribing detailed work processes and ensuring compliance with them than with continuous learning.

The Role of Power

Another recurrent theme is the role of power and power differences. The relationship, described in the previous sections, between the 'planners' and the 'doers' of work processes obviously involves differences of power. The formal work procedures forming the backbone of a safety management system can themselves be viewed as instruments of power. For instance, the existence of rules governing behaviour is sometimes used to attribute blame in the aftermath of accidents. Hopkins's (2000) analysis of the Longford gas explosion goes a long way to suggest that Esso deliberately tried to fix the responsibility of the accident on one particular operator in order to avoid any corporate responsibility and legal liability. In my data there are no stories about Norwegian oil companies employing such strategies, but the suspicion that procedures may serve to cover the backs of managers and companies is never far away when operational workers talk about them. This is a formal and 'systemic' kind of power, strongly linked to the recognized hierarchy of an organization. The case studies also identified more subtle and symbolically expressed examples of power.

In Chapter 4, I showed how the concepts of culture and power were inescapably intertwined. The case study in Chapter 5 illustrated empirically the connections between culture and power. The asymmetries of power and status identified in the relationship between the seamen on the supply vessels and the representatives of the oil and gas installations were almost entirely implicit and tacit. There were few examples of situations where the seamen were plainly and openly pressured by employees at the installations. And, of course, there were no explicit statements or documents specifying that the supply vessels were to be considered subordinate to the installations. In other words, the asymmetries of power and status between these two groups are almost entirely expressed symbolically, and as such would be very hard to identify without using a cultural approach. Even though they were only expressed symbolically, the implicit pressures and expectations that the vessels were to satisfy the whims of the oil company and the representatives of the installations were perceived to be very real by the seamen on the vessels. This serves as an example of why studies of the relationship between culture and safety should never disregard issues of power.

The Need for Ethnographically Inspired Methods

In addition to the balancing between centralization and decentralization and the interconnections between culture and power, arguments for more qualitative and ethnographically inspired methods are also a recurring theme in the book. This is partly due to the fact that all the empirical studies presented are based on a constructivist perspective on risk and safety. Also, the arguments for using ethnographically inspired methods are founded on the empirical findings from the study of Snorre Alpha, which exemplified some fundamental weaknesses with survey methods in regards to the study of organizational culture.

This argument may also be framed in a more general critique of organizational research. There is a striking difference between the early research on industrial organizations (e.g. Crozier 1964; Lysgaard 2001) and the more recent literature on organizational structure and design (e.g. Mintzberg 1983), leadership and management (e.g. Hersey and Blanchard 1977), organizational learning (e.g. Senge 1990; Argyris and Schön 1996), organizational culture (e.g. Schein 1992) or decision making (e.g. March 1994), to name a few. This difference lies in the degree of interest in the actual *work* that is performed within the boundaries of organizations. While the early industrial sociology was interested in the meanings that people attach to their work, the later research has been mainly preoccupied with processes and phenomena that are on a far higher level of abstraction, such as coordination, authority or workflow (Van Maanen and Barley 1984). By focusing on the abstract processes of organizing rather than real-life experiences they conveyed the implication that the way work is carried out and the way people relate to it were the unmentioned, well-kept secrets within the operational strata of organizations (Suchman 1995).

I suspect that the neglect of real-life experiences of work is the reason why the abstract concepts and nostrums of managers and organizational consultants, for example business process reengineering or total quality management, rarely live up to the expectations people have of them (Amundsen and Kongsvik 2008). This distance from operational work is a particular problem for safety research as it involves a risk of neglecting the sharp end of the organizations, i.e. the employees who are directly exposed to hazards. Ethnographically inspired methods can remedy some of this weakness since such methods explicitly aim at understanding the everyday practices in a field (Hammersley and Atkinson 1995). The descriptions and interpretations of everyday work practice which can be produced by such research may be instrumental for the management of safety because it should provide safety managers with a higher level of insight into the work processes they are managing and the context in which these work processes are situated.

Broadening the Concept of Safety Culture

It has been an objective of this book to elaborate the theoretical and methodological foundations for the study of safety culture. This has been done by studying cultural influences on safety in quite different contexts. While being a part of the same business and industry, the cases varied greatly with regard to the tasks being performed, and the hazards involved. The variation between the cases studied required a conception of culture that was broad enough to be able to relate to both complex, organizational accidents such as Snorre Alpha and occupational accidents with more limited consequences such as those on service vessels or supply bases. This requires a concept of culture that spans different levels of analysis from the 'macro' aspects, covering the organizations' ability to detect, correct and learn from mistakes, to the micro aspects, relating to the conventions for behaviour, interaction and communication between the members of work communities. These two aspects, while on different levels of analysis and abstraction, are of course highly interrelated.

One of the general 'products' of the book is therefore a broad conceptualization of culture which combines the informational perspective – culture as frames of reference – of Turner (1978) and Weick (1987; Weick et al. 1999), with a more practice-based conception – culture as conventions for behaviour, interaction and communication – inspired by the works of Gherardi and Nicolini (1998, 2000).

The cultural traits relevant for analyzing organizational accidents will probably have more to do with culture as frames of reference than culture as conventions for behaviour, interaction and communication. The opposite will apply to occupational accidents. Of course, both types of accident will be likely to involve both the dimensions of culture described here but the mix will be different in each case.

This is a visualization of culture which facilitates both an integrative view, i.e. the organization as a whole, and a differentiation perspective that recognizes the possible differences between the various communities of practice (Wenger 1998) which make up an organization.

Culture, Management and Learning – Stimulating the Growth of a Generative Organization

One of the reasons for the growing interest in the concept of safety culture is probably that some cultural traits are presumed to favour organizational learning with regard to safety. The organizational taxonomy proposed by Westrum (1993) has been particularly influential in this respect. His concept of a 'generative' organization, the characteristics of which include the active seeking of information, the training and rewarding of messengers and whistle-blowers and openness towards new ideas, has become the formulation of the ultimate goal of safety management. This book would be incomplete if it did not address the issue of organizational learning and how the research presented can contribute to reaching this elusive goal, so I shall offer a brief discussion of this issue.

First of all, the findings of this book have highlighted the role of employee participation in learning and improvement. Reducing the distance between managers and the operational workforce is vital for learning with regard to both facilitating the upward flow of information and the adaption of general safety measures to local work contexts.

Arguing for increased worker involvement does not mean, however, that workers and managers cannot have conflicting interests, or even that organizations should necessarily strive to circulate one particular culture or worldview. As I argued in Chapter 4, having multiple cultures and different interpretations of reality may serve as a form of requisite variety (cf. Westrum 1993; Weick 2001) that may increase the organization's ability to learn. In a recent article, Nævestad (2008) makes an interesting link between the concept of requisite variety and organizations' ability to recognize hazards. He argues that having multiple frames of reference within the boundaries of an organization serves as a 'cultural redundancy' that may reduce the organization's insensibility to hazards. Although Nævestad implicitly assumes the existence of some sort of overarching frame of reference – integration – in order to be able to communicate perceptions of risk across subcultural boundaries – differentiation – his article highlights some important cultural preconditions for organizational learning. The most important of these is that learning will be difficult in organizations where there is little variety in the way risks are interpreted and evaluated.

In order to make use of the requisite variety which exists in the different cultural frames of reference, the organization must be characterized by a high degree of trust and openness. This, in turn, presupposes the willingness of the powerful to 'set aside' their formal authority and let the power of argument take precedence over the power of position, in a way akin to Habermas's (1990) concept of the ideal speech situation.[3] The importance of such trust and openness has been one of the most central themes in safety culture research, for instance in Reason's conceptual framework where the fostering of a 'just culture' (Reason 1997: 205) is emphasized as a key prerequisite for learning (see also Dekker 2007). This trust is something that the managers of organizations must earn through consistent action over time. Studies of the relationship between organizational properties and levels of intra-organizational trust has shown that worker involvement (Gillespie and Mann 2004) and perceived procedural justice (Connell et al. 2003) are among the properties that seem to foster trust between workers and management. These findings resonate well with those of this book. The creation of a climate for learning seems to be a question of achieving a high degree of correspondence between the 'front of stage' slogans and safety visions, and the 'backstage' priorities in real decisions and situations.[4] Safety mantras like 'there is always time to work safely', which are not followed up by action and priorities conveying the same message, may therefore do more harm than good to the organization's learning abilities, by eroding the foundations for trust.

The general conclusion of the book is that a cultural approach to the study of safety can be highly useful in order to understand the match between formal and informal aspects of work and organizing. Culture influences safety on two levels. One, by constituting the frames of reference through which risks are recognized, evaluated, or altogether ignored. Two, culture influences safety by involving conventions for behaviour, interaction and communication.

I have argued for expanding the scope of the research on the cultural influences on safety on a theoretical, methodological and practical level.

On a theoretical level there is a need for a broadening of the analytical frames used to study safety culture. Particularly, cultural aspects should not be studied in isolation from other aspects of organizing, such as formal structures and processes of interaction. Also the notion of power needs to be included in the study of culture

---

3   The ideal speech situation is the core notion in Habermas's discourse ethics, and refers to a democratic ideal where there are no systematic distortions to communication. Grossly simplified, an ideal speech situation can be described as a situation where every participant in the discourse has equal rights to make statements, claims and counterclaims, interpretations and corrections, without reference to anything other than the strength of the argument (cf. Habermas 1990).

4   This use of Goffman's (1959) terms 'frontstage' and 'backstage' is borrowed from Langåker (2002).

in order to avoid the risk of painting too harmonious and integrative picture of organizational life.

On a methodological level there is a need for studies utilizing more ethnographically inspired methods. As culture is symbolically expressed and may also be largely taken for granted by the members of a cultural unit, the study of culture cannot rely on standardized assessment tools such as questionnaires. Rather, the study of culture requires interactive probing and sensitivity to the local context of work.

On a practical level there is a need for changing the way one conceives of the relationship between organizational culture and change. Culture should not be seen as the primary target of change efforts. Rather, knowledge of cultural traits and processes ought to be regarded as a crucial component for changing the *practices* taking place within the boundaries of an organization. The principles of action research have proved useful in this respect.

Many of the conclusions here might seem trivial to organizational researchers. The concept of organizational culture on which the work is based, the arguments towards using more ethnographically inspired methods for assessments as well as the arguments towards analyzing the interplay between cultural, structural and relational aspects of organization, are all based on existing knowledge. As such, the arguments in this book may not qualify for any awards in originality within the field of organizational research. But within that field the findings and arguments are not trivial. Safety research is a field where the insights of organizational research and theory have only just started to make their presence felt. Therefore there is a need for studies that seek to penetrate further how the knowledge of organizational science can be utilized in the study of safety.

References

ACSNI Human Factors Study Group (1993), *Organising for Safety* (Norwich, HSE Books).

Adler, R. B., Rosenfeld, L. B. and Towne, N. (1983), *Interplay. The Process of Interpersonal Communication* (New York, Holt, Rinehart and Winston).

Almklov, P. (2005), Radio – eller noen tanker om persepsjon, tanke og kultur (Radio – or some Thoughts about Perception, Thought and Culture) *Anthropology and Onthology* (S.E.Johansen, Trondheim, NTNU).

Alvesson, M. (2002), *Understanding Organizational Culture* (London, Sage).

Alvesson, M. and Sveningsson, S. (2008), *Changing Organizational Culture: Cultural Change Work in Progress* (London, Routledge).

Amundsen, O. (2009), Organisasjonens Små Og Store Fortellinger. Narrative Intervjuer Som Redskap I Studiet Av Organisasjonskultur (the Small and Big Stories of an Organization. Narrative Interviews as a Tool for the Study of Organizational Culture), in Hepsø, I. L. and Kongsvik, T. Ø. (eds) *Forskning Som Endringsverktøy I Organisasjoner: Forståelse Og Utvikling Av Praksis* (Trondheim, Tapir).

Amundsen, O. and Kongsvik, T. Ø. (2008), *Endringskynisme (Change Cynisism)* (Oslo, Gyldendal).

Andersen, F. S. (2001), *Den Meningsskapte Organisasjon (The Meaning-Created Organization)* (Oslo, Universitetsforlaget).

Antonsen, S. (2006), *Forenkling Av Styrende Dokumentasjon (Simplification of Governing Documents)*, (Trondheim, Studio Apertura).

Antonsen, S. (2009), Safety Culture and the Issue of Power. *Safety Science,* 47, 183–191.

Antonsen, S. (forthcoming), The Relationship between Culture and Safety on Offshore Supply Vessels. *Safety Science,* In Press, Corrected Proof.

Antonsen, S., Almklov, P. and Fenstad, J. (2008), Reducing the Gap between Procedures and Practice – Lessons from a Successful Safety Intervention. *Safety Science Monitor,* 12.

Antonsen, S. and Kongsvik, T. (2004), *Sikkerhetskultur Blant Deltakere I "Kollegaprogrammet for Bedre Sikkerhet" (Safety Culture among Participants of the Safe Behaviour Programme)*, (Trondheim, Studio Apertura).

Antonsen, S., Ramstad L.S. and Kongsvik, T. (2007), Unlocking the Organization: Action Research as a Means of Improving Organizational Safety. *Safety Science Monitor,* 11.

Argyris, C. and Schön, D. A. (1996), *Organizational Learning: Theory, Method, and Practice* (Reading, MA., Addison-Wesley).

Arner, O. (1961), *Skipet Og Sjømannen: Sosiologiske Undersøkelser Av Skipssamfunnet Og Sjøfolks Yrkesforhold (the Ship and the Seaman: Sociological Investigations of Ship Society and Seamen's Working Conditions)* (Oslo, Institutt for Samfunnsforskning).

Aubert, V. and Arner, O. (1962), *The Ship as a Social System* (Oslo, UiO/ISO).

Aven, T. and Kristensen, V. (2005), Perspectives on Risk: Review and Discussion of the Basis for Establishing a Unified and Holistic Approach. *Reliability Engineering & System Safety*, 90, 1–14.

Bachrach, P. and Baratz, M. S. (1962), Two Faces of Power. *American Political Science Review*, 56, 947–952.

Baker, J. (2007), *The Report of the BP US Refineries Independent Safety Review Panel* (Washington DC., US Chemical Safety and Hazard Investigation Board).

Barling, J. and Hutchinson, I. (2000), Commitment vs. Control-Based Safety Practices, Safety Reputation, and Perceived Safety Climate. *Canadian Journal of Administrative Sciences*, 14, 76–84.

Bax, E. H., Stejn, B. J. and De Witte, M. C. (1998), Risk Management at the Shop Floor: The Perception of Formal Rules in High-Risk Work Situations. *Journal of Contingencies and Crisis Management*, 6, 177.

Beck, U. (1992), *Risk Society* (London, Sage).

Berger, P. L. and Luckmann, T. (1966), *The Social Construction of Reality. A Treatise in the Sociology of Knowledge* (London, Penguin Books).

Bolman, L. G. and Deal, T. E. (2003), *Reframing Organizations. Artistry Choice and Leadership* (San Fransisco, Jossey-Bass).

Boudreau, F. A. and Newman, W. M. (1993), *Understanding Social Life: An Introduction to Sociology* (Minneapolis, West Publishing Company).

Bourrier, M. (1998), Elements for Designing a Self-Correcting Organisation: Examples from Nuclear Plants, in Hale, A. R. and Baram, M. S. (eds) *Safety Management: The Challenge of Change* (Oxford, Pergamon).

Bourrier, M. (2005), The Contribution of Organizational Design to Safety. *European Management Journal*, 23, 98–104.

Brattbakk, M., Østvold, L.-Ø., Van Der Zwaag, C. and Hiim, H. (2005), *Gransking Av Gassutblåsning På Snorre a, Brønn 34/7-P31 a 28.11.2004 (Investigation of Gas Blowout on Snorre Alpha, Well 34/7-P31 a 28.11.2004)* (Stavanger, Petroleum Safety Authority Norway).

Brunsson, N. (2000), *The Irrational Organization* (Bergen, Fagbokforlaget).

Bye, R., Antonsen, S. and Vikland, K. M. (2008), "Us" And "Them": The Impact of Group Identity on Safety Critical Behaviour, in Martorell, S., Guedes Soares, C. and Barnett, J. (eds) *Safety, Reliability and Risk Analysis: Theory, Methods and Applications*. Proceedings of the European Safety and Reliability Conference, Esrel 2008, and 17thsra-Europe, Valencia, Spain, September, 22–25, 2008 (Boca Raton, CRC Press).

Bye, R., Kongsvik, T. and Hansson, L. (2003), Establishing a Safety Culture in a Distributed Offshore Logistics Activity, in Bedford, T. and Gelder, P. H. A. J. M. V. (eds) *Safety and Reliability: Proceedings of Esrel 2003, European Safety and Reliability Conference 2003, 15–18 June 2003, Maastricht, the Netherlands* (Lisse, Balkema Publishers).

Castells, M. (2001), *The Rise of the Network Society* (Oxford, Blackwell).

Clarke, L. B. (2006), *Worst Cases: Terror and Catastrophe in the Popular Imagination* (Chicago, University of Chicago Press).

Clarke, S. (1999), Perceptions of Organizational Safety: Implications for the Development of Safety Culture. *Journal of Organizational Behavior,* 20, 185–198.

Clegg, S. R. (1989), *Frameworks of Power* (London, Sage).

Columbia Accident Investigation Board (CAIB) (2003), *Final Report on Columbia Space Shuttle Accident* (Washington D.C, U.S. Government Printing Office).

Connell, J., Ferres, N. and Travaglione, T. (2003), Engendering Trust in Manager-Subordinate Relationships: Predictors and Outcomes. *Personnel Review,* 32, 569–587.

Cooper, M. D. and Phillips, R. A. (2004), Exploratory Analysis of the Safety Climate and Safety Behavior Relationship. *Journal of Safety Research,* 35, 497–512.

Cox, S. and Flin, R. (1998), Safety Culture: Philosopher's Stone or Man of Straw? *Work Stress,* 12, 189–201.

Cox, S. J. and Cheyne, A. J. T. (2000), Assessing Safety Culture in Offshore Environments. *Safety Science,* 34, 111–129.

Crozier, M. (1964), *The Bureaucratic Phenomenon* (Chicago, University of Chicago Press).

Cullen, D. (1990), *The Public Inquiry into the Piper Alpha Disaster* (London, HMSO).

Cyert, R. M. and March, J. G. (1963), *A Behavioral Theory of the Firm* (Englewood Cliffs, N.J., Prentice-Hall).

Czarniawska, B. (1997), *Narrating the Organization: Dramas of Institutional Identity* (Chicago, University of Chicago Press).

Dagbladet (2005), *Kaster Opp Fem, Seks Ganger Hver Natt (Throws up Five, Six Times Per Night).*

Dahl, R. A. (1957), The Concept of Power. *Behavioral Science,* 201–215.

Dale-Olsen, H. (2005), *Det Nye Arbeidsmarkedet: Kunnskapsstatus Og Problemstillinger (the New Working Life: Knowledge Status and Problems to Be Addressed)* (Oslo, Norges forskningsråd, Program for arbeidslivsforskning).

Dallner, M. (2000), *Validation of the General Nordic Questionnaire (Qpsnordic) for Psychological and Social Factors at Work* (København, Nordisk Ministerråd).

Deal, T. E. and Kennedy, A. A. (1982), *Corporate Cultures: The Rites and Rituals of Corporate Life* (Reading, Mass., Addison-Wesley).

DeJoy, D. M. (2005), Behavior Change Versus Culture Change: Divergent Approaches to Managing Workplace Safety. *Safety Science,* 43, 105–129.

Dekker, S. (2006), Resilience Engineering: Chronicling the Emergence of Confused Consensus, in Hollnagel, E., Woods, D. D. and Leveson, N. (eds) *Resilience Engineering - Concepts and Precepts* (Aldershot, Ashgate).

Dien, Y. (1998), Safety and Application of Procedures, or 'How Do "They" Have to Use Operating Procedures in Nuclear Power Plants'. *Safety Science,* 29, 179–187.

Douglas, M. (1992), *Risk and Blame: Essays in Cultural Theory* (London, Routledge).

Douglas, M. and Wildavsky, A. (1982), *Risk and Culture: An Essay on the Selection of Technical and Environmental Dangers* (Berkeley, California, University of California Press).

Drottz-Sjøberg, B.-M. (2003), *Current Trends in Risk Communication* (Oslo, Directorate for Civil Defence and Emergency Planning).

Duijm, N. J. and Goossens, L. (2006), Quantifying the Influence of Safety Management on the Reliability of Safety Barriers. *Journal of Hazardous Materials,* 130, 284–292.

Eckstein, H. (1975), Case Study and Theory in Political Science, in Greenstein, I. and Polsby, N. W. (eds) *Handbook of Political Science* (Reading, MA, Addison-Wesley).

Ek, A., Akselsson, R., Arvidsson, M. and Johansson, C. R. (2007), Safety Culture in Swedish Air Traffic Control. *Safety Science,* 45, 791–811.

Emery, M. (1999), *Searching: The Theory and Practice of Making Cultural Change* (Amsterdam, John Benjamins).

Eriksen, T. H. (1998), *Små Steder – Store Spørsmål: Innføring I Sosialantropologi (Small Places – Big Questions: Introduction to Social Anthropology)* (Oslo, Universitetsforlaget).

Farrington-Darby, T., Pickup, L. and Wilson, J. R. (2005), Safety Culture in Railway Maintenance. *Safety Science,* 43, 39–60.

Feldman, M. S. (1991), *The Meanings of Ambiguity, Reframing Organizational Culture* (Newbury Park, Sage).

Fenstad, J., Kongsvik, T., Solem, A. and Antonsen, S. (2007), *Sikkerhet Og Arbeidsmiljø På Statoils Servicefartøyer (Safety and Work Environment on Statoil's Service Vessels),* (Trondheim, NTNU Samfunnsforskning AS, Studio Apertura).

Fernández-Muñiz, B., Montes-Peón, J. M. and Vázquez-Ordás, C. J. (2007), Safety Culture: Analysis of the Causal Relationships between Its Key Dimensions. *Journal of Safety Research,* 38, 627–641.

Fischhoff, B. (1975), Hindsight Is Not Equal to Foresight: The Effect of Outcome Knowledge on Judgment under Uncertainty. *Journal of Experimental Psychology: Human Perception and Performance,* 1, 288–299.

Fischhoff, B., Lichtenstein, S., Slovic, P., Derby, S. L. and Keeney, R. L. (1981), *Acceptable Risk* (Cambridge, Cambridge University Press).

Flin, N., Mearns, K., O'connor, P. and Bryden, R. (2000), Measuring Safety Climate: Identifying the Common Features. *Safety Science,* 34, 177–192.

Flin, R. (2007), Measuring Safety Culture in Healthcare: A Case for Accurate Diagnosis. *Safety Science,* 45, 653–667.

Florczak, C. M. (2002), *Maximizing Profitability with Safety Culture Development* (Amsterdam, Butterworth-Heinemann).

Foucault, M. (1977), *Discipline and Punish: The Birth of the Prison* (London, Allen Lane).

Foucault, M. and Gordon, C. (1980), *Power/Knowledge: Selected Interviews and Other Writings 1972–1977* (Brighton, Harvester Press).

Frost, P. J., Moore, L. F., Louis, M. R., Lundberg, C. C. and Martin, J. (1991), *Reframing Organizational Culture* (Newbury Park, Sage).

Geertz, C. (1973), *The Interpretation of Cultures: Selected Essays* (New York, Basic Books).

Gherardi, S. and Nicolini, D. (2000), The Organizational Learning of Safety in Communities of Practise. *Journal of Management Inquiry,* 9, 7–18.

Gherardi, S., Nicolini, D. and Odella, F. (1998), Toward a Social Understanding of How People Learn in Organizations. *Management Learning,* 29, 273–297.

Gherardi, S., Nicolini, D. and Odella, F. (1998), What Do You Mean by Safety? Conflicting Perspectives on Accident Causation and Safety Management in a Construction Firm. *Journal of Contingencies and Crisis Management,* 6, 202–213.

Giddens, A. (1984), *The Constitution of Society: Outline of the Theory of Structuration* (Cambridge, Polity Press).

Giddens, A. (1994), *Sociology* (Cambridge, Polity Press).

Gillespie, N. A. and Mann, L. (2004), Transformational Leadership and Shared Values: The Building Blocks of Trust. *Journal of Managerial Psychology,* 19, 588–607.

Glaser, B. G. and Strauss, A. L. (1967), *The Discovery of Grounded Theory* (New York, Aldine De Gruyter).

Glendon, A. I. and Stanton, N. A. (2000), Perspectives on Safety Culture. *Safety Science,* 34, 193–214.

Goffman, E. (1959), *The Presentation of Self in Everyday Life* (Garden City, N.Y., Doubleday).

Goffman, E. (1961), *Asylums: Essays on the Social Situation of Mental Patients and Other Inmates* (New York, Doubleday).

Goodenough, W. (1994), Toward a Working Theory of Culture, in Borofsky, R. (ed.) *Assessing Cultural Anthropology* (New York, McGraw-Hill).

Gouldner, A. W. (1954), *Patterns of Industrial Bureaucracy* (New York, Free Press).

Granovetter, M. (1973), The Strength of Weak Ties. *American Journal of Sociology,* 78.

Greenwood, D. J. and Levin, M. (1998), *Introduction to Action Research: Social Research for Social Change* (Thousand Oaks, California, Sage Publications).

Gregory, K. L. (1983), Native-View Paradigms: Multiple Cultures and Culture Conflicts in Organizations. *Administrative Science Quarterly,* 28, 359–376.

Grote, G. (2007), Understanding and Assessing Safety Culture through the Lens of Organizational Management of Uncertainty. *Safety Science,* 45, 637–652.

Grote, G. (2008), Diagnosis of Safety Culture: A Replication and Extension Towards Assessing "Safe" Organizational Change Processes. *Safety Science,* 46, 450–460.

Guldenmund, F. W. (2000), The Nature of Safety Culture: A Review of Theory and Research. *Safety Science,* 34, 215–257.

Guldenmund, F. W. (2006), Much Ado About Safety Culture. *Safety Science Monitor,* 11.

Guldenmund, F. W. (2007), The Use of Questionnaires in Safety Culture Research - an Evaluation. *Safety Science,* 45, 723–743.

Habermas, J. (1990), *Moral Consciousness and Communicative Action* (Cambridge, Polity Press).

Hale, A. (2000), Editorial: Culture's Confusions. *Safety Science,* 34, 1–14.

Hale, A., Heijer, F. and Koornneeff, F. (2003), Management of Safety Rules: The Case of Railways. *Safety Science Monitor,* 1, 1–11.

Hale, A. and Hovden, J. (1998), Management and Culture: The Third Age of Safety. A Review of Approaches to Organizational Aspects of Safety, Health and Environment, in Feyer, A.-M. (ed.) *Occupational Injury: A Selection of Papers Presented at the Occupational Injury Symposium, Sydney, Australia, February 1996* (Exeter, Pergamon).

Hale, A. R. (1990), Safety Rules Ok? Possibilities and Limitations in Behavioral Safety Strategies. *Journal of Occupational Accidents,* 12, 3–20.

Hammersley, M. and Atkinson, P. (1995), *Ethnography: Principles in Practice* (London, Routledge).

Hansson, L., Antonsen, S., Solem, A., Marthinsen, S., Kongsvik, T., Ellefsen, M. and Sætre, G. (2004), *Evaluering Og Prioritering Av Sikkerhetstiltak.(Evaluation and Prioritization of Safety Measures)* (Trondheim, Studio Apertura, NTNU).

Haukelid, K. (1999), *Risiko Og Sikkerhet (Risk and Safety)* (Oslo, Universitetsforlaget).

Haukelid, K. (2008), Theories of (Safety) Culture Revisited – An Anthropological Approach. *Safety Science,* 46, 413–426.

Heath, R. (1998), Looking for Answers: Suggestions for Improving How We Evaluate Crisis Management. *Safety Science,* 30, 151–163.

Heinrich, H. W. (1931), *Industrial Accident Prevention: A Scientific Approach* (New York, McGraw-Hill).

Helmreich, R. L. (2000), On Error Management: Lessons from Aviation. *British Medical Journal,* 320, 781–785.

Hersey, P. and Blanchard, K. H. (1993), *Management of Organizational Behavior: Utilizing Human Resources* (Englewood Cliffs, N.J., Prentice Hall).

Hidden, A. (1989), *Investigation into the Clapham Junction Railway Accident* (London, HMSO).

Hinds, P. and Kiesler, S. (2002), *Distributed Work* (Cambridge, Mass., MIT Press).

Hofmann, D. and Mark, B. (2006), An Investigation of the Relationship between Safety Climate and Medication Errors as Well as Other Nurse and Patient Outcomes. *Personnel Psychology,* 59, 847–869.

Hofstede, G. (1984), *Culture's Consequences* (Newbury Park, Sage).

Hollnagel, E. and Woods, D. D. (2006), Epilogue: Resilience Engineering Precepts, in Hollnagel, E., Woods, D. D. and Leveson, N. (eds) *Resilience Engineering - Concepts and Precepts* (Aldershot, Ashgate).

Hollnagel, E., Woods, D. D. and Leveson, N. (2006), Prologue: Resilience Engineering Concepts, in Hollnagel, E. (ed.) *Resilience Engineering - Concepts and Precepts* (Aldershot, Ashgate).

Hollnagel, E., Woods, D. D. and Leveson, N. (2006), *Resilience Engineering - Concepts and Precepts* (Aldershot, Ashgate).

Hopkins, A. (2000), *Lessons from Longford: The Esso Gas Plant Explosion* (Sydney, CCH Australia Ltd).

Hopkins, A. (2005), *Safety, Culture and Risk. The Organisational Causes of Disasters* (Sydney, CCH Australia Ltd.).

Hopkins, A. (2006), Studying Organisational Cultures and Their Effects on Safety. *Safety Science,* 44, 875–889.

Hovden, J. (1991), *Safety Climate and Culture: The Basic Elements*, Mo-SHE-course, Module 7.

Hovden, J. and Larsson, T. J. (1987), Risk: Culture and Concepts, in Singleton, W. T. and Hovden, J. (eds) *Risk and Decisions* (Chichester, John Wiley & Sons).

Health and Safety Executive (HSE) (2001), Health and Safety Climate Tool (Norwich, Health and Safety Executive).

Huang, Y. H., Ho, M., Smith, G. S. and Chen, P. Y. (2006), Safety Climate and Self-Reported Injury: Assessing the Mediating Role of Employee Safety Control. *Accident Analysis & Prevention,* 38, 425–433.

Hudson, P., Parker, D. and Van De Graaf, G. C. (2002), The Hearts and Minds Program: Understanding Hse Culture, *Proceedings of the 6th Spe International Conference on Health Safety and Environment in Oil and Gas Exploration and Production.* (Richardson, Society of Petroleum Engineers).

Hudson, P. T. W., Verschuur, W. L. G., Parker, D., Lawton, R., Van Der Graaf, G. and Kalff, J. (no date), *Bending the Rules: Violation in the Workplace*, Retrieved March 20, 2007, from http://www.energyinst.org.uk/heartsandminds/docs/bending.pdf.

Hviid Nielsen, T. (1994), *Risici - I Teknologien, I Samfundet Og I Hovederne (Risks - in Technology, in Society and in People's Minds)* (Oslo, TMV-senteret).

Höpfl, H. (1994), Safety Culture, Corporate Culture. Organizational Transformation and the Commitment to Safety. *Disaster Prevention and Management,* 3, 49–58.

Håvold, J. I. (2007), *From Safety Culture to Safety Orientation. Developing a Tool to Measure Safety in Shipping* (Trondheim, Norwegian University of Science and Technology).

Håvold, J. I. and Nesset, E. (2009), From Safety Culture to Safety Orientation: Validation and Simplification of a Safety Orientation Scale Using a Sample of Seafarers Working for Norwegian Ship Owners. *Safety Science, 47*, 305–326.

International Atomic Energy Agency (IAEA) (1986), *Summary Report on the Post-Accident Review Meeting on the Chernobyl Accident* (Wien, IAEA).

International Atomic Energy Agency (IAEA (1991), *Safety Culture. International Nuclear Safety Advisory Group Safety Series 75, Insag-4* (Wien, IAEA).

International Atomic Energy Agency (IAEA (1992), *The Chernobyl Accident. International Nuclear Safety Advisory Group Safety Series Updating of Insag-1 : Insag-7* (Wien, IAEA).

Janis, I. L. (1982), *Groupthink: Psychological Studies of Policy Decisions and Fiascoes* (Boston, Houghton Mifflin).

Johnson, C. W. (1996), Integrating Human Factors and Systems Engineering to Reduce the Risk of Operator "Error". *Safety Science, 22*, 195–214.

Jovchelovitch, S. and Bauer, M. W. (2000), Narrative Interviewing, in Bauer, M. W. and Gaskell, G. (eds) *Qualitative Researching with Text, Image and Sound: A Practical Handbook* (London, Sage).

Justis- Og Politidepartementet (2008), St.Meld. Nr. 22 (2007–2008) Samfunnssikkerhet - Samvirke Og Samordning.

Kalleberg, R. (1985), Kombinering Av Forskningstradisjoner I Sosiologien (the Combination of Research Traditions in Sociology), in Dale, B., Jones, M. and Martinussen, W. (eds) *Metode På Tvers: Samfunnsvitenskapelige Forskningsstrategier Som Kombinerer Metoder Og Analysenivåer* (Trondheim, Tapir).

Karwal, A. K., Verkaik, R. and Jansen, C. (2000), *Non-Adherence to Procedures - Why Does It Happen* (EASS, Paper presented at the 12th annual European Aviation Safety Seminar).

Keesing, R. M. (1987), Anthropology as Interpretive Quest. *Current Anthropology, 28*, 161–176.

Keesing, R. M. (1994), Theories of Culture Revisited, in Borofsky, R. (ed.) *Assessing Cultural Anthropology* (New York, McGraw-Hill).

Knudsen, F. (2005), Sømandskab. Arbejdsidentitet Og Sikkerhedsregler Hos Danske Sømænd (Seamanship. Work Identity and Safety Rules of Danish Seamen). *Tidsskriftet Antropologi.*

Kongsvik, T. (2003), Hvilke Barrierer? Ansattes Vurdering Av Sider Ved Sikkerhetskulturen – Snorre a (Which Barriers? Employees' Perception of Aspects of Safety Culture - Snorre Alpha), (Trondheim, Studio Apertura).

Krause, T. R., Hidley, J. H. and Hodson, S. J. (1990), *The Behavior-Based Safety Process: Managing Involvement for an Injury-Free Culture* (New York, Van Nostrand Reinhold).

Kringen, J. (forthcoming), *Culture and Control. Regulation of Risk in the Norwegian Petroleum Industry* (Oslo, University of Oslo).

Kroeber, A. L. and Kluckhohn, C. (1963), *Culture: A Critical Review of Concepts and Definitions* (New York, Vintage Books).

Lamvik, G. M. (2002), *The Filipino Seafarer: A Life between Sacrifice and Shopping*, (Trondheim, Dept. of Social Anthropology, Norwegian University of Science and Technology).

Langåker, L. (2002), *Reframing Organisational Safety: A Multiperspective Cultural Approach* (Bath, University of Bath).

Laporte, T. R. and Consolini, P. M. (1991), Working in Practise but Not in Theory. *Journal of Public Administration and Theory,* 1, 19–47.

Laurence, D. (2005), Safety Rules and Regulations on Mine Sites - the Problem and a Solution. *Journal of Safety Research,* 36, 39–50.

Law, J. (1991), Power, Discretion and Strategy, *A Sociology of Monsters: Essays on Power, Technology and Domination* (London, Routledge).

Lawrie, M., Parker, D. and Hudson, P. (2006), Investigating Employee Perceptions of a Framework of Safety Culture Maturity. *Safety Science,* 44, 259–276.

Lawton, R. (1998), Not Working to Rule: Understanding Procedural Violations at Work. *Safety Science,* 28, 77–95.

Lecompte, M. D. and Schensul, J. J. (1999), *Designing & Conducting Ethnographic Research* (Walnut Creek, Calif., AltaMira Press).

Lee, T. and Harrison, K. (2000), Assessing Safety Culture in Nuclear Power Stations. *Safety Science,* 34, 61–97.

Lee, T. R. (1996), Perceptions, Attitudes and Behaviour: The Vital Elements of a Safety Culture. *Health and Safety,* October, 1–15.

Levin, M. and Klev, R. (2002), *Forandring Som Praksis: Læring Og Utvikling I Organisasjoner (Change as Practice: Learning and Development in Organizations)* (Bergen, Fagbokforlaget).

LeVine, R. A. (1984), Properties of Culture. An Ethnographic View, in Shweder, R. A. and Levine, R. A. (eds) *Culture Theory. Essays on Mind, Self, and Emotion* (New York, Cambridge University Press).

Lin, S.-H., Tang, W.-J., Miao, J.-Y., Wang, Z.-M. and Wang, P.-X. (2008), Safety Climate Measurement at Workplace in China: A Validity and Reliability Assessment. *Safety Science,* 46, 1037–1046.

Lind, M. (1979), *The Use of Flow Models for Design of Plant Operating Procedures* (Garching, Germany, IWG/NPPCI specialists' meeting on procedures and systems for assisting an operator during normal and anomalous nuclear power plant operation situations).

Ljung, M. and Jenstad, T. E. (1987), *Banning I Norsk, Svensk Og 18 Andre Språk (Cursing in Norwegian, Swedish and 18 Other Languages)* (Oslo, Universitetsforlaget).

Lukes, S. (1974), *Power: A Radical View* (London, Macmillan).

Lukes, S. (2005), *Power: A Radical View* (Basingstoke, Palgrave Macmillan).

Lupton, D. (1999), *Risk* (London, Routledge).

Lysgaard, S. (2001), *Arbeiderkollektivet (The Workers' Collective)* (Oslo, Universitetsforlaget).

Madsen, M. D. (2006), Improving Patient Safety: Safety Culture and Patient Safety Ethics, *Risø National Laboratory.* (Roskilde, University of Roskilde).

Malinowski, B. (1922), *Argonauts of the Western Pacific: An Account of Native Enterprise and Adventure in the Archipelagoes of Melanesian New Guinea* (London, Routledge & Kegan Paul).

March, J. G. (1994), *A Primer on Decision Making: How Decisions Happen* (New York, Free Press).

March, J. G., Olsen, J. P. and Christensen, S. (1976), *Ambiguity and Choice in Organizations* (Bergen, Universitetsforlaget).

Martin, J. (1992), *Cultures in Organizations: Three Perspectives* (New York, Oxford University Press).

Martin, J. (2002), *Organizational Culture: Mapping the Terrain* (Thousand Oaks, Calif., Sage Publications).

Martin, J. and Meyerson, D. (1988), Organizational Cultures and the Denial, Channeling and Acknowledgment of Ambiguity, *Managing Ambiguity and Change* (Chichester, John Wiley & Sons).

Martin, J. and Siehl, C. (1983), Organizational Culture and Counterculture. *Organizational Dynamics,* Autumn.

Mason, S. (1997), Procedural Violations – Causes, Costs and Cures, *Human Factors in Safety-Critical Systems* (Oxford, Butterworth-Heinemann).

Mead, M. (1926), *Coming of Age in Samoa: A Psychological Study of Primitive Youth for Western Civilisation* (New York, William Morrow & Company).

Mearns, K., Whitaker, S. M. and Flin, R. (2003), Safety Climate, Safety Management Practice and Safety Performance in Offshore Environments. *Safety Science,* 41, 641–680.

Medvedev, G. (1991), *The Truth About Chernobyl,* (New York, Basic Books).

Meshkati, N. (1996), *A Major Cause of the Chernobyl Accident: The Vital Role of Human and Organizational Factors in the Safety of Nuclear Power Plants* (Paris, May 22–24, 1996, A Brief Invited Position Paper for the Conference From Chernobyl to Nuclear Safety in Eastern Europe and Newly Independent States: The Search for a New Partnership).

Meyer, J. W. and Rowan, B. (1977), Institutionalized Organizations – Formal-Structure as Myth and Ceremony. *American Journal of Sociology,* 83, 340–363.

Meyerson, D. (1991), Acknowledging and Uncovering Ambiguities in Cultures, *Reframing Organizational Culture* (Newbury Park, Sage).

Meyerson, D. and Martin, J. (1987), Cultural Change: An Integration of Three Different Views. *Journal of Management Studies,* 24, 623–647.

Miles, M. B. and Huberman, A. M. (1994), *Qualitative Data Analysis: An Expanded Sourcebook* (Thousand Oaks, Calif., Sage).

Mintzberg, H. (1983), *Structure in Fives: Designing Effective Organizations* (Englewood, New Jersey, Prentice Hall).

Mitchison, N. and Papadakis, G. A. (1999), Safety Management Systems under Seveso II: Implementation and Assessment. *Journal of Loss Prevention in the Process Industries,* 12, 43–51.

Morgan, G. (1986), *Images of Organization* (Beverly Hills, CA, Sage).

Nisbett, R. E. and Wilson, T. D. (1977), Telling More Than We Can Know: Verbal Reports on Mental Processes. *Psychological Review,* 84, 231–251.

Nævestad, T.-O. (2008), Safety Cultural Preconditions for Organizational Learning in High-Risk Organizations. *Journal of Contingencies and Crisis Management,* 16, 154–163.

Ogbonna, E. and Harris, L. C. (1998), Managing Organizational Culture: Compliance or Genuine Change? *British Journal of Management,* 9, 273–288.

Ogbonna, E. and Harris, L. C. (2005), The Adoption and Use of Information Technology: A Longitudinal Study of a Mature Family Firm. *New Technology, Work & Employment,* 20, 2–18.

Ogbonna, E. and Harris, L. C. (2006), Organisational Culture in the Age of the Internet: An Exploratory Study. *New Technology, Work & Employment,* 21, 162–175.

Ogbonna, E. and Wilkinson, B. (2003), The False Promise of Organizational Culture Change: A Case Study of Middle Managers in Grocery Retailing. *Journal of Management Studies,* 40, 1151–1178.

Olive, C., O'Connor, T. M. and Mannan, M. S. (2006), Relationship of Safety Culture and Process Safety. *Journal of Hazardous Materials,* 130, 133–140.

Parker, D., Lawrie, M. and Hudson, P. (2006), A Framework for Understanding the Development of Organisational Safety Culture. *Safety Science,* 44, 551–562.

Parker, G. M. (2006), What Makes a Team Effective or Ineffective, in Gallos, J. V. (ed.) *Organization Development* (San Francisco, Josey-Bass).

Perrow, C. (1984), *Normal Accidents* (New York, Basic Books).

Perrow, C. (1994), Accidents in High-Risk Systems. *Technology Studies,* 1, 1–38.

Perrow, C. (1999), *Normal Accidents. 2. Edition* (Princeton, Princeton University Press).

Peters, T. and Waterman, R. H. (1982), *In Search of Excellence* (New York, Harper & Row).

Petroleum Safety Authority (PSA) (2005), *Trends in Risk Levels on the Norwegian Continental Shelf - Phase 6 Summary.* (Stavanger, Petroleum Safety Authority Norway).

Pfeffer, J. (1981), *Power in Organisations* (Cambridge, MA, Ballinger Publishing).

Pfeffer, J. (1992), *Managing with Power* (Boston, Harvard Business School Press).

Pidgeon, N. (1997), The Limits to Safety? Culture, Politics, Learning and Man-Made Disasters. *Journal of Contingencies and Crisis Management,* 5, 1–14.

Pidgeon, N. (1998), Safety Culture: Key Theoretical Issues. *Work and Stress,* 12, 202–216.

Pidgeon, N. and O'Leary, M. (2000), Man-Made Disasters: Why Technology and Organizations (Sometimes) Fail. *Safety Science,* 34, 15–30.

Pidgeon, N. F., Turner, B. A., Blockley, D. I. and Toft, B. (1991), Corporate Safety Culture: Improving the Management Contribution to System Reliability, in Matthews, R. H. (ed.) *"International Conference on Reliability Techniques and their Application, Reliability '91", 10–12 June 1991, London.* (London, Elsevier).

Porter, M. (1985), *The Value Chain and Competitive Advantage* (New York, Free Press).

Rasmussen, J. (1997), Risk Management in a Dynamic Society: A Modeling Problem. *Safety Science,* 27, 183–213.

Reason, J. (1987), The Chernobyl Errors. *Bulletin of the British Psychological Society,* 40, 201–206.

Reason, J. (1990), *Human Error* (Cambridge, Cambridge University Press).

Reason, J. (1997), *Managing the Risks of Organisational Accidents* (Aldershot, Ashgate).

Reason, J., Parker, D., Lawton, R. and Pollock, C. (1995), *Organisational Controls and the Varieties of Rule-Based Behaviour* (London, Risk in Organisational Settings Conference, ESRC Risk and Human Behaviour Programme).

Reason, P. and Bradbury, H. (2008), *The Sage Handbook of Action Research: Participative Inquiry and Practice* (London, SAGE).

Reichers, A. E. and Schneider, B. (1990), Climate and Culture: An Evolution of Constructs, in Schneider, B. (ed.) *Organizational Climate and Culture* (San Francisco, Jossey-Bass).

Richter, A. (2001), *Nye Ledelsesformer, Sikkerhedskultur Og Forebyggelse Af Ulykker. Hovedrapport (New Forms of Leadership, Safety Culture and the Prevention of Accidents)* (Lyngby, Danmarks Tekniske Universitet).

Richter, A. (2003), New Ways of Managing Prevention: A Cultural and Participative Approach. *Safety Science Monitor.*

Richter, A. and Koch, C. (2004), Integration, Differentiation and Ambiguity in Safety Cultures. *Safety Science,* 42, 703–722.

Roberts, K. H. (1993), *New Challenges to Understanding Organizations* (New York, Macmillan).

Rochlin, G. I. (1993), Defining 'High Reliability' Organization in Practice; a Taxonomic Prologue, in Roberts, K. H. (ed.) *New Challenges to Understanding Organizations* (New York, Macmillan).

Rochlin, G. I., La Porte, T. M. and Roberts, K. H. (1987), The Self-Designing High-Reliability Organization: Aircraft Carrier Flight Operations at Sea. *Naval War College Review,* 40, 76–90.

Rogers, W. P. and Et, A. (1986), *Report of the Presidential Commission on the Space Shuttle Challenger Accident* (Washington D.C, U.S Government Printing Office).

Rosness, R. (2001), Safety Culture - Just Another Buzzword to Hide Our Confusion?, unpublished paper. Available online at http://www.risikoforsk.no/Publikasjoner/Safety%20culture.pdf.

Roughton, J. E. and Mercurio, J. (2002), *Developing an Effective Safety Culture: A Leadership Approach* (Boston, Butterworth-Heinemann).

Rousseau, D. (1990), *Assessing Organizational Culture: The Case for Multiple Methods, Organizational Culture and Climate* (San Fransisco, Jossey-Bass).

Rundmo, T. (2000), Safety Climate, Attitudes and Risk Perception in Norsk Hydro. *Safety Science,* 34, 47–59.

Ryggvik, H. (2003), *Fra Forvitring Til Ny Giv - Om En Storulykke Som Aldri Inntraff? (from Disintegration to New Initiative. On a Major Accident That Never Happened)* (Oslo, TIK-Center).

Ryggvik, H. (2008), *Adferd, Teknologi Og System: En Sikkerhetshistorie (Behaviour, Technology and System: A Safety History)* (Trondheim, Tapir).

Safety Climate Assessment Toolkit (SCAT) (no date), *Safety Climate Measurement: User Guide and Toolkit.* (Loughborough, Loughborough University Business School). Available online at http://www.lboro.ac.uk/departments/bs/safety/document.pdf.

Sagan, S. D. (1993), *The Limits of Safety* (Princeton, Princeton University Press).

Salvendy, G. (2006), *Handbook of Human Factors and Ergonomics* (Hoboken, N.J., Wiley).

Schein, E. H. (1992), *Organizational Culture and Leadership* (San Fransisco, Jossey-Bass).

Schein, E. H. (2003), *DEC Is Dead, Long Live DEC: The Lasting Legacy of Digital Equipment Corporation* (San Francisco, Berrett-Koehler).

Schein, E. H. (2004), Letters to the Editor. Reply To: In Defence of Schein's Perspective on Organizational Culture. *Safety Science,* 42, 980–981.

Schiefloe, P. M. (1977), *Miljø Om Bord (The Social Environment Onboard)* (Trondheim, SINTEF).

Schiefloe, P. M. (2003), *Mennesker Og Samfunn: Innføring I Sosiologisk Forståelse (Man and Society: Introduction to Sociological Understanding)* (Bergen, Fagbokforlaget).

Schiefloe, P. M., Vikland, K., Ytredal, E. B., Torsteinsbø, A., Moldskred, I., Heggen, S., Sleire, D. H., Førsund, S. A. and Syversen, J. E. (2005), *Årsaksanalyse Etter Snorre a-Hendelsen 28.11.2004 (Causal Investigation into the Snorre Alpha Incident 28.11.2004)*, Statoil).

Schneider, B. (1990), *Organizational Climate and Culture* (San Francisco, Jossey-Bass).

Scholte, B. (1984), Comment On: The Thick and the Thin: On the Interpretive Theoretical Program of Clifford Geertz, by P.Shankman. *Current Anthropology,* 25, 261–270.

Schön, D. A. (1987), *Educating the Reflective Practitioner* (San Francisco, Jossey-Bass).

Sejersted, F. (1993), *Managere Og Konsulenter Som Manipulatorer. Noen Refleksjoner Om Etikkens Suspensjon (Managers and Consultants as Manipulators. Some Reflections on the Suspension of Ethics)* (TMV, Senter for teknologi og menneskelige verdier).

Senge, P. M. (1990), *The Fifth Discipline: The Art and Practice of the Learning Organization* (New York, Doubleday).

Shannon, H. S., Robson, L. S. and Guastello, S. J. (1999), Methodological Criteria for Evaluating Occupational Safety Intervention Research. *Safety Science,* 31, 161–179.

Shore, B. (1996), *Culture in Mind: Cognition, Culture, and the Problem of Meaning* (New York, Oxford University Press).

Shrader-Frechette, K. S. (1991), *Risk and Rationality : Philosophical Foundations for Populist Reforms* (Berkeley, University of California Press).

Slovic, P., Fischhoff, B. and Lichtenstein, S. (1980), Facts and Fears: Understanding Perceived Risk, in Schwing, R. C. and Albers, W. A. (eds) *Societal Risk Assessment: How Safe Is Safe Enough?* (New York, Plenum Press).

Smircich, L. (1983), Concepts of Culture and Organizational Analysis. *Administrative Science Quarterly,* 28, 339–358.

Snook, S. A. (2000), *Friendly Fire: The Accidental Shootdown of U.S. Black Hawks over Northern Iraq* (Princeton, Princeton University Press).

Suchman, L. (1995), Making Work Visible. *Communications of the ACM,* 38, 56–64.

Suchman, L. A. (1987), *Plans and Situated Actions: The Problem of Human-Machine Communication* (Cambridge, Cambridge University Press).

Swidler, A. (1986), Culture in Action – Symbols and Strategies. *American Sociological Review,* 51, 273–286.

Sætre, P. O. (2006), *Improvisasjon Som Middel Til Vellykket Gjenvinning Av Situasjonskontroll (Improvization as a Means for Successful Recovery of Situation Control)* (Trondheim, Norwegian University of Science and Technology).

Thagaard, T. (2003), *Systematikk Og Innlevelse: En Innføring I Kvalitativ Metode (Systematics and Insight: An Introduction to Qualitative Methods)* (Bergen, Fagbokforlaget).

Tharaldsen, J. E. and Haukelid, K. (2007), Culture or Behaviour? Friend or Foe?, in Aven, T. and Vinnem, J. E. (eds) *Risk Reliability and Societal Safety* (London, Taylor and Francis).

Tharaldsen, J. E., Olsen, E. and Rundmo, T. (2007), A Longitudinal Study of Safety Climate on the Norwegian Continental Shelf. *Safety Science,* In Press, Corrected Proof.

Tharaldsen, J. E., Olsen, E. and Rundmo, T. (2008), A Longitudinal Study of Safety Climate on the Norwegian Continental Shelf. *Safety Science,* 46, 427–439.

Tinmannsvik, R. K. (2008), "Stille Avvik" - Trussel Eller Mulighet? ("Silent Deviations" - Threat or Possibility?), in Tinmannsvik, R. K. (ed.) *Robust Arbeidspraksis: Hvorfor Skjer Det Ikke Flere Ulykker På Sokkelen? (Robust Work Practice: Why Do'nt More Accidents Occur on the Continental Shelf?)* (Trondheim, Tapir).

Todnem, G. (2009), Hvordan Kan Forskning Bidra I Endringsprosser? (How Can Research Contribute in Change Processes?), in Hepsø, I. L. and Kongsvik, T. (eds) *Forskning Som Endringsverktøy (Research as a Tool for Change)* (Trondheim, Tapir).

Turner, B. (1978), *Man-Made Disasters* (London, Wykenham Science Press).

Turner, B. A. (1991), The Development of a Safety Culture. *Chemistry and Industry,* 1, 241–243.

Turner, B. A. (1992), *The Sociology of Safety, Engineering Safety* (London, McGraw-Hill).

Turner, B. A. and Pidgeon, N. F. (1997), *Man-Made Disasters. 2nd Edition* (Oxford, Butterworth Heinemann).

Tversky, A. and Kahneman, D. (1974), Judgment under Uncertainty: Heuristics and Biases. *Science,* 185, 1124–1131.

Tylor, E. B. (1968), The Science of Culture, in Fried, M. (ed.) *Readings in Anthropology, Vol.2 Cultural Anthropology* (New York, Crowell).

Van Maanen, J. (1991), The Smile Factory: Work at Disneyland, in Frost, P. J., Moore, L. F., Louis, M. R., Lundberg, C. C. and Martin, J. (eds) *Reframing Organizational Culture* (Newbury Park, Sage).

Van Maanen, J. and Barley, S. R. (1984), Occupational Communities. Culture and Control in Organizations, in Staw, B. M. and Cummings, L. L. (eds) *Research in Organizational Behavior* (Greenwich, JAI Press Ltd.).

Van Maanen, J. and Barley, S. R. (1985), Cultural Organization: Fragments of a Theory, *Organizational Culture* (Newbury Park, CA, Sage Publications).

Vaughan, D. (1996), *The Challenger Launch Decision* (Chicago, The University of Chicago Press).

Vaughan, D. (1997), The Trickle-Down Effect: Policy Decisions, Risky Work, and the Challenger Tragedy. *California Management Review,* 39, 80–102.

Weber, M. (1971), *Makt Og Byråkrati (Power and Bureaucracy)* (Oslo, Gyldendal).

Weick, K. (1987), Organizational Culture as a Source of High Reliability. *California Management Review,* 29, 112–127.

Weick, K. (1991), The Vulnerable System: An Analysis of the Tenerife Air Disaster, in Frost, P. J., Moore, L. F., Louis, M. R., Lundberg, C. C. and Martin, J. (eds) *Reframing Organizational Culture* (Newbury Park, Sage).

Weick, K., Sutcliffe, K. M. and Obstfeld, D. (1999), Organizing for High Reliability: Processes of Collective Mindfulness. *Research in Organizational Behavior,* 21, 81–123.

Weick, K. E. (1995), *Sensemaking in Organizations* (Thousand Oaks, Calif., Sage).

Weick, K. E. (2001), *Making Sense of the Organization* (Oxford, Blackwell).

Weick, K. E. and Sutcliffe, K. M. (2007), *Managing the Unexpected: Resilient Performance in an Age of Uncertainty* (San Francisco, Calif., Jossey-Bass).

Weisbord, M. (2006), Designing Work: Structure and Process for Learning and Self-Control, in Gallos, J. V. (ed.) *Organization Development* (San Francisco, Jossey-Bass).

Wenger, E. (1998), *Communities of Practice: Learning, Meaning, and Identity* (Cambridge, Cambridge University Press).

Westrum, R. (1993), Cultures with Requisite Imagination, in Wise, J. A., Hopkin, V. D. and Stager, P. (eds) *Verification and Validation of Complex Systems* (Berlin, Springer).

Whitehead, A. N. (1929), *Process and Reality: An Essay in Cosmology: Gifford Lectures Delivered in the University of Edinburgh During the Session 1927–28.* (New York, Macmillan).

Whyte, W. F. (1943), *Street Corner Society: The Social Structure of an Italian Slum* (Chicago, University of Chicago Press).

Wickens, C. D., Gordon-Becker, S. and Liu, Y. (2004), *An Introduction to Human Factors Engineering* (Upper Saddle River, N.J., Pearson Prentice Hall).

Wildavsky, A. (1988), *Searching for Safety* (New Brunswick, N.J., Transaction Books).

Wilkins, A. L. and Ouchi, W. G. (1983), Efficient Cultures: Exploring the Relationship between Culture and Organizational Performance. *Administrative Science Quarterly,* 28, 468–481.

Wittgenstein, L. (1997), *Filosofiske Undersøkelser (Philosophical Investigations)* (Oslo, Pax).

Wolf, E. (1994), Facing Power: Old Insights, New Questions, *Assessing Cultural Anthropology* (New York, McGraw-Hill).

Woods, D. (2003), *How Resilience Engineering Can Transform NASA's Approach to Risky Decision Making*, Testimony on The Future of NASA for Committee on Commerce, Science and Transportation).

Zerubavel, E. (1981), *Hidden Rhythms: Schedules and Calendars in Social Life* (Chicago, University of Chicago Press).

Zohar, D. (1980), Safety Climate in Industrial Organizations: Theoretical and Applied Implications. *Journal of Applied Psychology,* 65, 96–102.

Zohar, D. (2000), A Group-Level Model of Safety Climate: Testing the Effect of Group Climate on Microaccidents in Manufacturing Jobs. *Journal of Applied Psychology,* 85, 587–596.

Index

Name Index